Series of Evaluation Reports on National Marine Innovation

国家海洋创新评估系列报告

Guojia Haiyang Chuangxin Pinggu Xilie Baogao

National Marine Innovation Index Report 2017-2018 (English Chinese Revision)

国家海洋创新指数报告 2017～2018（英汉修订版）

— Liu Dahai　He Guangshun　Wang Chunjuan —

（刘大海　何广顺　王春娟）/著

科学出版社

北京

内 容 简 介

本报告以海洋创新数据为基础，构建了国家海洋创新指数指标体系，客观分析了我国海洋创新现状与发展趋势，定量评估了国家和区域海洋创新能力，探讨了我国海洋科研机构的空间分布特征与演化趋势，并对我国海洋创新能力进行了评价与展望。同时，对比分析了全球海洋创新能力，并开展了国际海洋科技研究态势和我国海洋国家实验室等专题分析。

本报告既适用于海洋领域的专业科技工作者和研究生、大学生，也是海洋管理和决策部门的重要参考资料，还可为全社会认识和了解我国海洋创新发展提供窗口。

Based on marine innovation data, this report establishes national marine innovation index system to objectively analyze the current status and development trends of China's marine innovation and quantitatively assess national and regional marine innovation capabilities. In addition, it discusses spatial distribution characteristics and evolutionary trends of marine scientific research institutions in China and evaluates and looks ahead to marine innovation capabilities of China. Finally, this report makes comparative analysis of global marine innovation capabilities and conducts specific analysis of the international marine scientific and technological research situations and China's marine national laboratory.

This report is useful for both scientific and technological practioner in the marine field and university students, either graduate or undergraduate ones. It is not only an important reference book for marine management and decision-making departments but also acts as a window for the whole society to gain an understanding of marine innovation development in China.

图书在版编目(CIP)数据

国家海洋创新指数报告 2017-2018（英汉修订版）: National Marine Innovation Index Report 2017-2018 (English Chinese Revision) / 刘大海，何广顺，王春娟著. —北京：科学出版社，2019.11

（国家海洋创新评估系列报告）

ISBN 978-7-03-062778-0

Ⅰ. ①国… Ⅱ. ①刘… ②何… ③王… Ⅲ. ①海洋经济-技术革新-研究报告-中国-2017-2018-英文 Ⅳ. ①P74

中国版本图书馆 CIP 数据核字（2019）第 233104 号

责任编辑：朱 瑾 田明霞 / 责任校对：郑金红
责任印制：吴兆东 / 封面设计：无极书装

科学出版社 出版
北京东黄城根北街 16 号
邮政编码：100717
http://www.sciencep.com

北京虎彩文化传播有限公司 印刷
科学出版社发行 各地新华书店经销

*

2019 年 11 月第 一 版　　开本：889×1194　1/16
2019 年 11 月第一次印刷　　印张：13 1/2
字数：372 000

定价：180.00 元

（如有印装质量问题，我社负责调换）

National Marine Innovation Index Report 2017-2018 (English Chinese Revision) Academic Committee

Director: Li Tiegang

Committee members: Xuan Zhaohui Gao Feng Gao Runsheng
Pan Kehou Zhu Yingchun Li Renjie
Xu Xingyong Wang Xiao

Consultants: Ding Dewen Jin Xianglong Wu Lixin Qu Tanzhou
Xin Hongmei Wang Xiaoqiang Qin Haoyuan Ma Deyi
Yu Xingguang Wei Zexun Wang Zongling Lei Bo
Zhang Wen Wen Quan Shi Xuefa Wang Baodong
Feng Lei Wang Yuan

Writers: Liu Dahai He Guangshun

Compilation group members: Liu Dahai Wang Chunjuan Xu Meng
Xiang Jiahao Wang Jinping Li Xianjie
Yin Xigang Lu Jingliang Yu Ying
Liu Weifeng Lin Xianghong
Liang Chenjing Wang Qi Wang Xiqian
Wang Xiyuan Zhao Qian

Calculation group members: Wang Chunjuan Xu Meng Xiang Jiahao
Wang Qi Wang Xiyuan Wang Xiqian

Translators: Yang Hong Sun Guangfeng Xu Xin

Author Affiliations: First Institute of Oceanography, Ministry of Natural Resources
National Marine Data and Information Service
Lanzhou Information Center, Chinese Academy of Sciences
Pilot National Laboratory for Marine Science and Technology(Qingdao)

《国家海洋创新指数报告2017~2018(英汉修订版)》学术委员会

主　　任：李铁刚

委　　员：玄兆辉　高　峰　高润生　潘克厚　朱迎春
　　　　　李人杰　徐兴永　王　骁

顾　　问：丁德文　金翔龙　吴立新　曲探宙　辛红梅
　　　　　王孝强　秦浩源　马德毅　余兴光　魏泽勋
　　　　　王宗灵　雷　波　张　文　温　泉　石学法
　　　　　王保栋　冯　磊　王　源

著　　者：刘大海　何广顺

撰写组：刘大海　王春娟　徐　孟　项佳皓　王金平
　　　　李先杰　尹希刚　鲁景亮　于　莹　刘伟峰
　　　　林香红　梁琛婧　王　琦　王玺茜　王玺媛
　　　　赵　倩

测算组：王春娟　徐　孟　项佳皓　王　琦　王玺媛
　　　　王玺茜

翻译组：杨　红　孙广峰　许　新

著者单位：自然资源部第一海洋研究所
　　　　　国家海洋信息中心
　　　　　中国科学院兰州文献情报中心
　　　　　青岛海洋科学与技术试点国家实验室

Preface

The report of the 19th National Congress of the Communist Party of China put forward that "innovation is the primary driving force behind development" and "we will improve our national innovation system and boost our strategic scientific and technological strength". The 13th Five-Year Plan period is the deciding stage for building China into a well-off society in an all-encompassing manner and the key stage of implementing an innovation-driven development strategy and building maritime power. Marine innovation is an important component of national innovation, and also the primary driving source for the realization of a maritime power. The report of the 19th National Congress of the Communist Party of China stated that "We will implement the coordinated regional development strategy, pursue coordinated land and marine development, and step up efforts to build China into a strong maritime country. We should put the Belt and Road Initiative on top priority, give equal emphasis to 'bringing in' and 'going out', increase openness and cooperation in building innovation capacity. With these efforts, we hope to make new ground in opening China further through links running eastward and westward, across land and over sea".

In response to the national marine innovation strategy, and for the sake of serving the building of a national innovation system, the First Institute of Oceanography, Ministry of Natural Resources (MNR), set about implementing the measurement and calculation of marine innovation indicators beginning in 2006, and initiated the research on national marine innovation indexes in 2013. With the help and support of relevant departments, seven volumes of National Marine Innovation Index Reports have been published since 2015. *National Marine Innovation Index Report 2017-2018* is the eighth one.

On the basis of data such as marine economic statistics, science and technology statistics, and technological achievements registration, the *National Marine Innovation Index Report 2017-2018* establishes the index system for national marine innovation indexes in four aspects: marine innovation resources, marine knowledge creation, marine innovation performance and marine innovation environment. National marine innovation indexes from 2004 to 2016 have been quantitatively measured and assessed from the four aspects. This report makes an objective analysis of both national and regional marine innovation capacities. It also analyzes the characteristics and evolutionary trend of the spatial distribution of marine scientific research institutions, comparatively analyzes global marine innovation capacities, and makes a special analysis for the research on international marine science and technology and marine national laboratory, effectively reflecting the quality and efficiency of marine innovation in China.

National Marine Innovation Index Report 2017-2018 was compiled by the Marine Policy Research Center of the First Institute of Oceanography, MNR. Lanzhou Information Center, Chinese Academy of Sciences co-wrote the sections about marine papers, marine patents, global marine innovation capabilities and the conditions of international marine science and technology research. Pilot National Laboratory for Marine Science and Technology (Qingdao) did a special analysis about marine national laboratories.

National Marine Data and Information Service, Department of Inovation and Development, Ministry of Science and Technology of the People 's Republic of China, Science and Technology Department of the Ministry of Education and other units and departments provided data support. Chinese Academy of Science and Technology for Development offered technical support in the evaluation system, measurement, and calculation methods. Here sincere thanks go to all organizations and individuals who contributed to their compilation and data and technology support.

We hope that this series of evaluation reports on the national marine innovation index can act as a window for the entire society to gain an understanding of marine innovation development in China. Since this report is the staged achievement of research in the national marine innovation index, we earnestly accept the criticisms and suggestions from our colleagues if there are problems or deficiencies. Any valuable opinions of experts and scholars are welcome to help us keep improving the series of evaluation reports on the national marine innovation index.

Liu Dahai, He Guangshun
July 2018

前 言

党的十九大报告指出"创新是引领发展的第一动力",要"加强国家创新体系建设,强化战略科技力量"。"十三五"时期是我国全面建成小康社会的决胜阶段,是实施创新驱动发展战略、建设海洋强国的关键时期。海洋创新是国家创新的重要组成部分,也是实现海洋强国战略的动力源泉。十九大报告同时提出,"实施区域协调发展战略""坚持陆海统筹,加快建设海洋强国""要以'一带一路'建设为重点,坚持引进来和走出去并重""加强创新能力开放合作,形成陆海内外联动、东西双向互济的开放格局"。

为响应国家海洋创新战略、服务国家创新体系建设,自然资源部第一海洋研究所自2006年着手开展海洋创新指标的测算工作,并于2013年启动国家海洋创新指数的研究工作。在有关部门的帮助和支持下,国家海洋创新指数系列报告自 2015 年以来已经出版了七册,《国家海洋创新指数报告2017~2018》是该系列报告的第八册。

《国家海洋创新指数报告2017~2018》基于海洋经济统计、科技统计和科技成果登记等权威数据,从海洋创新资源、海洋知识创造、海洋创新绩效、海洋创新环境四个方面构建指标体系,定量测算2004~2016年我国海洋创新指数。客观评价了我国国家和区域海洋创新能力,研究了我国海洋科研机构的空间分布特征与演化趋势,对比分析了全球海洋创新能力,并对国际海洋科技研究态势和海洋国家实验室进行了专题分析,切实反映了我国海洋创新的质量和效率。

《国家海洋创新指数报告2017~2018》由自然资源部第一海洋研究所海洋政策研究中心组织编写。中国科学院兰州文献情报中心参与编写了海洋论文、海洋专利、全球海洋创新能力和国际海洋科技研究态势专题分析等部分,青岛海洋科学与技术试点国家实验室参与编写了海洋国家实验室专题分析部分。国家海洋信息中心、科学技术部战略规划司、教育部科学技术司等单位和部门提供了数据支持。中国科学技术发展战略研究院对评价体系与测算方法给予了技术支持。在此对参与编写和提供数据与技术支持的单位及个人,一并表示感谢。

希望国家海洋创新评估系列报告能够成为全社会认识和了解我国海洋创新发展的窗口。本报告是国家海洋创新指数研究的阶段性成果,如有问题或不足,敬请各位同仁批评指正,编写组会汲取各方面专家学者的宝贵意见,不断完善国家海洋创新指数系列报告。

刘大海 何广顺
2018 年 7 月

Contents

Preface

I. Viewing Marine Innovation in China from the Perspective of Data ···1
    1.1   Stable Structure of Human Resources for Marine Innovation ···3
    1.2   Gradual Improvement of Platform Environment for Marine Innovation ·································6
    1.3   Significant Increase in the Funding Scale of Marine Innovation ···8
    1.4   Continuous Growth of Marine Innovation Output Achievements ···11
    1.5   Steady Improvement in Marine Innovation of Higher Institutions ··23
    1.6   Steady Enhancement of Marine S&T Contributions to Marine Economic Development ···········30

II. Evaluation of National Marine Innovation Index ··32
    2.1   Comprehensive Evaluation of Marine Innovation Index ··34
    2.2   Evaluation of the Sub-index of Marine Innovation Resources ···37
    2.3   Evaluation of the Sub-index of Marine Knowledge Creation ··39
    2.4   Evaluation of the Sub-index of Marine Innovation Performance ···41
    2.5   Evaluation of the sub-index of Marine Innovation Environment ··43

III. Evaluation of Regional Marine Innovation Index ···45
    3.1   Viewing China's Regional Marine Innovation Development from the Perspective of Coastal
          Provinces (Autonomous Regions, Municipalities directly under the Central Government) ·········47
    3.2   Viewing China's Regional Marine Innovation Development from the Perspective
          of Five Economic Zones ···52
    3.3   Viewing China's Regional Marine Innovation Development from the Perspective
          of Three Marine Economic Circles ··54

IV. Progress and Prospect of China's Marine Innovation Capability ·······································56
    4.1   Complementation between National Marine Innovation Capability and Marine
          Economic Development ··57
    4.2   Progress of the Key Indicators Relevant to the 13th Five-Year Plan National
          Marine Planning ···58

**V. Spatial Distribution Characteristics and Evolutionary Trend of China's Marine
Scientific Research Institutions** ··59
    5.1   Research Methods ··60
    5.2   Spatial Distribution Characteristics and Evolutionary Trends ··60
    5.3   Major Research Conclusions ···67

VI. Analysis of Global Marine Innovation Capability ···69
    6.1   Analysis of Total Amount and Situation of Global Marine Innovation Achievements ···············71
    6.2   Comparative Analysis of National Strength ···76
    6.3   Analysis of Patented Technology Achievements in Marine Field ··79

Appendixes ··84
    Appendix 1 Indicator System of National Marine Innovation Index ································85
    Appendix 2 Indicator Interpretations of National Marine Innovation Index ····················88
    Appendix 3 Evaluation Methods of National Marine Innovation Index ·························90
    Appendix 4 Evaluation Methods of Regional Marine Innovation Index·························92
    Appendix 5 Calculation Methods of Contribution Rate of Marine S&T Progress···········93
    Appendix 6 Calculation Methods of Transformation Rate of Marine S&T Achievements·····················96
    Appendix 7 Regional Classification Basis and Definition of Related Concepts················98
    Appendix 8 List of Marine-related Higher Institutions (Including Marine-related Coefficient of Proportionality) ···99
    Appendix 9 List of Marine-related Disciplines (Discipline Classifications of the Ministry of Education)··100

Compilation Explanations ···104
Instructions on Updates··111

目　录

前言

第一章　从数据看我国海洋创新 ·· 112
　　第一节　海洋创新人力资源结构稳定 ··· 113
　　第二节　海洋创新平台环境逐渐改善 ··· 115
　　第三节　海洋创新经费规模显著提升 ··· 117
　　第四节　海洋创新产出成果持续增长 ··· 120
　　第五节　高等学校海洋创新稳步提升 ··· 131
　　第六节　海洋科技对经济发展贡献稳步增强 ·································· 136

第二章　国家海洋创新指数评价 ·· 138
　　第一节　海洋创新指数综合评价 ·· 139
　　第二节　海洋创新资源分指数评价 ·· 142
　　第三节　海洋知识创造分指数评价 ·· 143
　　第四节　海洋创新绩效分指数评价 ·· 145
　　第五节　海洋创新环境分指数评价 ·· 146

第三章　区域海洋创新指数评价 ·· 149
　　第一节　从沿海省（自治区、直辖市）看我国区域海洋创新发展 ········· 150
　　第二节　从五大经济区看我国区域海洋创新发展 ···························· 154
　　第三节　从三大海洋经济圈看我国区域海洋创新发展 ······················ 155

第四章　我国海洋创新能力的进步与展望 ·· 157
　　第一节　国家海洋创新能力与海洋经济发展相辅相成 ······················ 158
　　第二节　国家海洋"十三五"相关规划重要指标进展 ······················· 159

第五章　我国海洋科研机构的空间分布特征与演化趋势 ····················· 160
　　第一节　研究方法 ·· 161
　　第二节　空间分布特征与演化趋势 ·· 161
　　第三节　主要研究结论 ·· 167

第六章　全球海洋创新能力分析 ·· 168
　　第一节　全球海洋创新成果总量与态势分析 ·································· 169
　　第二节　国家实力对比分析 ·· 173
　　第三节　海洋领域专利技术成果分析 ··· 176

附录 ·· 180
　　附录一　国家海洋创新指数指标体系 ··· 181
　　附录二　国家海洋创新指数指标解释 ··· 183
　　附录三　国家海洋创新指数评价方法 ··· 185
　　附录四　区域海洋创新指数评价方法 ··· 186
　　附录五　海洋科技进步贡献率测算方法 ·· 187

附录六　海洋科技成果转化率测算方法 ··· 189
附录七　区域分类依据及相关概念界定 ··· 190
附录八　主要涉海高等学校清单(含涉海比例系数) ··· 191
附录九　涉海学科清单(教育部学科分类) ··· 191

编制说明 ··· 195

更新说明 ··· 200

I. Viewing Marine Innovation in China from the Perspective of Data

Under the strategic background of building a maritime power and the Belt and Road Initiative, China continuously makes new major achievements in marine innovation with some areas reaching the international advanced level, and both the conditions and environment for marine innovation are enhanced dramatically.

Human resources for marine innovation are constantly optimized. The structure of staff involved in scientific & technological (S&T) activities at marine research institutions is continuously improved; the overall number of research and development (R&D) staff along with the workload equivalent to full-time work rises steadily and the structure of academic qualifications of R&D personnel is further optimized.

National platforms for marine innovation remain stable. The number of national (key/engineering) laboratories of marine scientific research institutions and the number of national engineering (research/technology research) centers have maintained stable in recent years. The capital construction and the fixed assets of marine scientific research institutions increase year by year.

The funding scale of marine innovation significantly increases. The R&D funding scale of the marine scientific research institutions has significantly improved and the internal spending of R&D funding has also seen a steady growth.

Marine innovation achievements grow steadily. The total amount of marine S&T papers from the marine scientific research institutions maintains its momentum and the number of papers published and recorded on the Science Citation Index (SCI) in the marine field has grown substantially. The types of publications on marine science and technology have markedly increased, and the number of patent applications and grants shows robust growth.

Marine innovation at higher institutions steadily improves. Marine-related higher institutions witness year-on-year growth in personnel, funding, projects, and so on.

Marine science and technology make gradual and steady contributions to the development of the marine economy. The contribution rate of marine S&T progress and the transformation rate of marine S&T achievements in 2016 reached 65.9%[①] and 50.0%[②] respectively. Marine S&T innovation plays an increasingly prominent role in promoting achievement transformation.

[①] The contribution rate of marine S&T progress in 2016 was measured and calculated based on relevant data from 2006 to 2016
[②] The transformation rate of marine S&T achievements in 2016 was measured and calculated based on relevant data from 2000 to 2016

1.1 Stable Structure of Human Resources for Marine Innovation

Human resources for marine innovation are the main force and strategic resource for the building of a maritime power and an innovation-driven country. The overall quality of marine innovation researchers determines the speed and range in boosting national marine innovation capability. The personnel involved in S&T activities and R&D at marine research institutions are important human resources for marine innovation, reflecting the talent reserve status of human resources for national marine innovation. Among them, the personnel involved in S&T activities refer to the staff engaged in S&T activities at the marine scientific research institutions, including technological management staff, personnel involved in project activities, and technical service personnel. R&D personnel refer to the staff of the marine scientific research institutions and external researchers, and the postgraduate students who participate in R&D projects, R&D project management staff, and the staff providing direct services for R&D activities.

1.1.1 Continuous optimization of the structure of staff involved in S&T activities

With regard to personnel structure, from 2011 to 2016, the proportion of staff engaged in project activities at China's marine scientific research institutions (specifically the staff of the research office or research group) in staff involved in S&T activities has maintained above 64%, with 2016 witnessing a significant increase; both the S&T management personnel (namely the institution heads, business and management personnel) and technical service personnel (to be precise all types of personnel working directly for S&T services) have been almost maintaining below 15%, with 2014 and 2016 witnessing a relatively higher percentage (Figure 1-1). In terms of the academic qualification structure of staff, from 2011 to 2016, the proportion of doctoral and master graduates among the staff of China's marine scientific research institutions engaged in S&T activities has shown an overall upward trend. In 2016, the doctoral

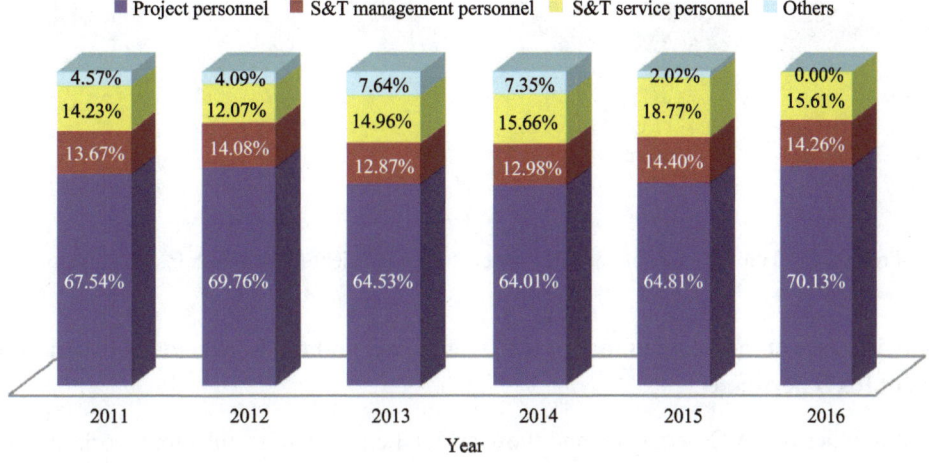

Figure 1-1 Composition of S&T Personnel at Marine Scientific Research Institutions from 2011 to 2016[①]

---

① The sum of the percentage of the data in the figure is not equal to 100% because some data have been rounded

and master graduates accounted for 28.14% and 32.32% respectively of the total personnel involved in S&T activities, with both categories witnessing an increase compared to 2015 (Figure 1-2). In terms of the professional title structure of the staff, from 2011 to 2016, the proportion of the S&T staff holding senior and intermediate professional titles in marine scientific research institutions has been obviously higher than that of staff with junior professional titles. In 2016, the staff with senior and intermediate professional titles accounted for 41.85% and 34.15% of the total S&T staff (Figure 1-3).

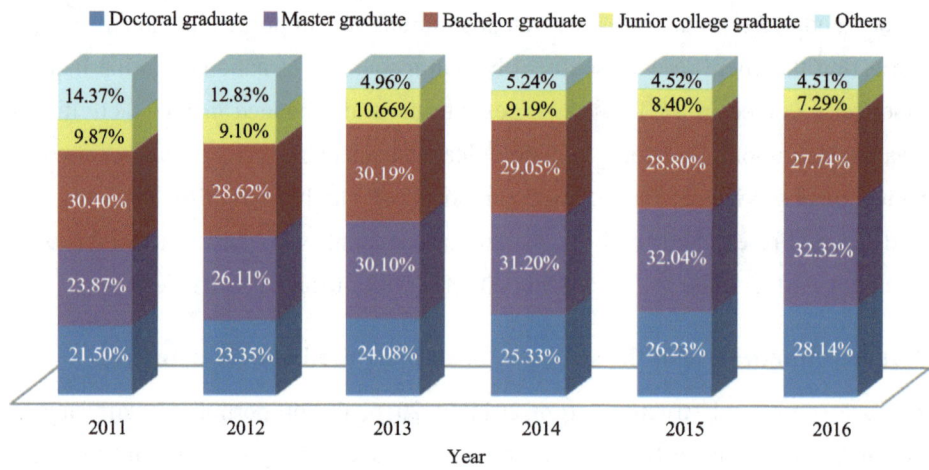

Figure 1-2　Academic Qualifications Structure of S&T Personnel at Marine Scientific Research Institutions from 2011 to 2016

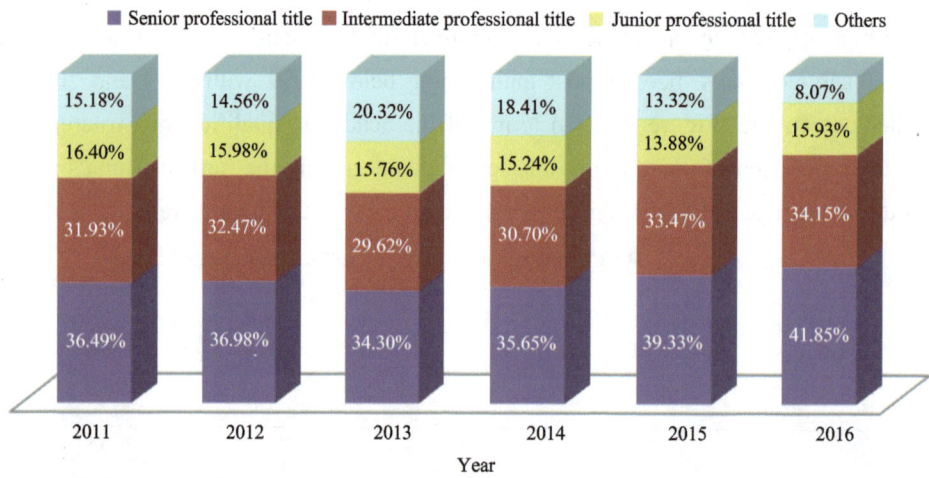

Figure 1-3　Professional Title Structure of S&T Personnel at Marine Scientific Research Institutions from 2011 to 2016

1.1.2　Steady increase in the total number of R&D personnel and the workload equivalent to full-time work of R&D personnel

The total number of R&D personnel and the workload equivalent to full-time work at China's marine scientific research institutions presented a steadily increasing trend (Figure 1-4). From 2002 to 2006, the increase was moderately slower; the years 2006 and 2007 witnessed a sharp increase, with a growth rate at 119.1% and 88.16% respectively; from 2007 to 2014, both presented a stable increase; from 2014 to 2015, the total number of R&D personnel witnessed a slight decline; from 2015 to 2016, both presented another significant increase, with a growth rate at 13.68% and 6.55% respectively.

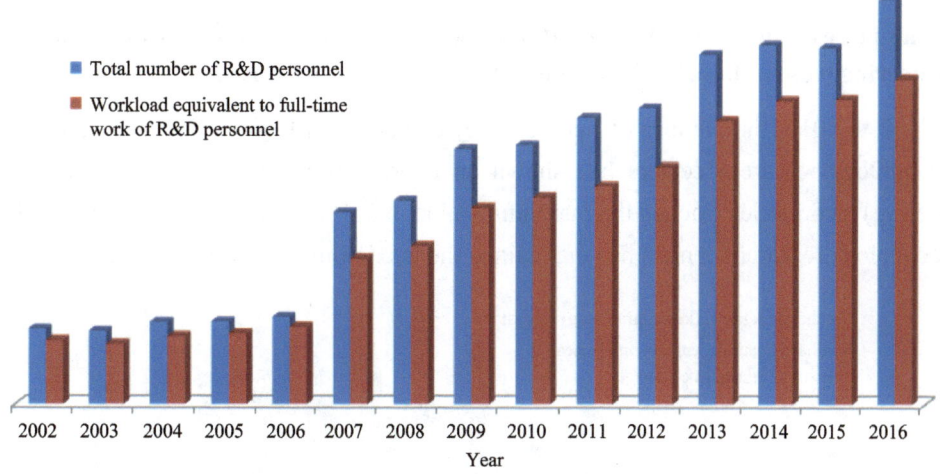

Figure 1-4　Total Number (Person) and the Workload Equivalent to Full-time Work (Person·Year) of R&D Personnel at Marine Scientific Research Institutions (Person) from 2002 to 2016

1.1.3　Stable structure of academic qualifications of R&D personnel

From 2011 to 2016, the number of doctoral graduates in R&D personnel from marine scientific research institutions has maintained an upward trend, but the proportion has shown an increasing trend in fluctuations. The number of master graduates presented an overall growth trend. In 2016, graduates with doctoral and master's degrees accounted for 31.67% and 32.97% of the total number of R&D personnel respectively (Figure 1-5). The number of doctoral graduates accounted for the highest proportion in 2015, reaching 31.99%, an increase of 4.17% compared to 2011. The proportion of master graduates has maintained a steady growth in fluctuations, up by 6.07% in 2016 over the year 2011.

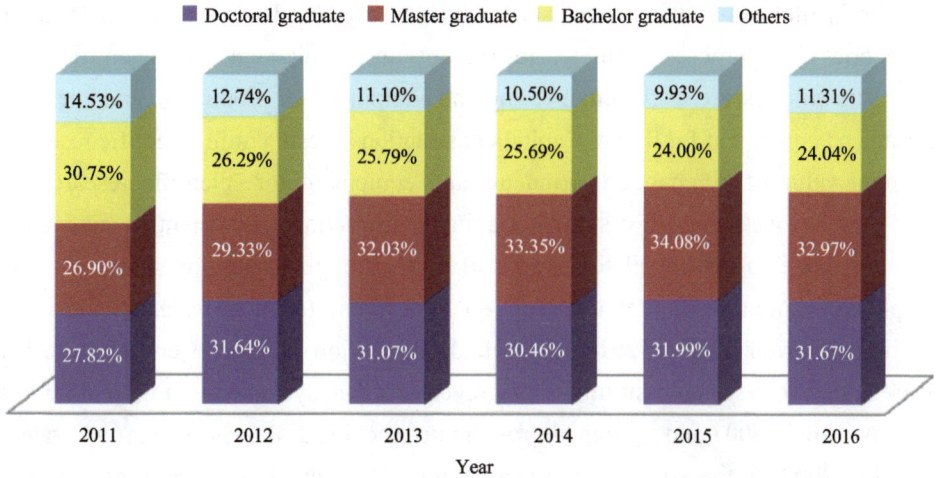

Figure 1-5　Academic Qualification Structure of R&D Personnel at Marine Scientific Research Institutions from 2011 to 2016

1.2 Gradual Improvement of Platform Environment for Marine Innovation

1.2.1 Overall increase in the number of national (key/engineering) laboratories and national engineering (research/technology research) centers of marine research institutions

From 2002 to 2016, the number of national (key/engineering) laboratories and national engineering (research/technology research) centers has shown an overall increasing trend, the number of national (key/engineering) laboratories reached the maximum value in 2010, and the number of national engineering (research/technology research) centers has maintained the maximum value from 2013 to 2015 (Figure 1-6).

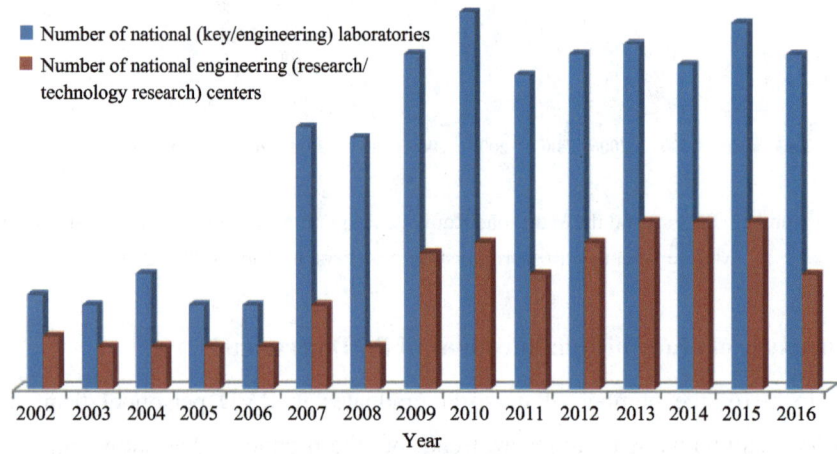

Figure 1-6　Number of National (Key/Engineering) Laboratories and National Engineering (Research/Technology Research) Centers of Marine Scientific Research Institutions from 2002 to 2016

1.2.2 Continuous increase in the actual completion of capital construction investment

The actual completion of capital construction investment refers to the capital construction workload completed by the institutions in the current year, which is expressed in currency. It is divided into the following categories by purposes, namely, instrument and equipment for scientific research, civil engineering for scientific research, civil engineering and equipment for production and operation as well as civil engineering and equipment for living. Capital construction investment in scientific research equipment refers to the total value of purchased instruments and equipment for scientific research in the actual completion of capital construction investment. Capital construction investment in civil engineering for scientific research refers to the completed workload of civil engineering for scientific research in the actually completed amount of capital construction investment (such as scientific research building, laboratory building). From 2002 to 2016, the actual completion of capital construction investment of China's marine scientific research institutions has revealed a steady increase (Figure 1-7) with the most rapid growth appearing in 2007 and an annual growth rate reaching 228.66%. The growth rate of 2016 was 28.66 times that of 2002. In the aspect of use classification, the actual completion of capital construction investment from 2002 to 2016 was mainly in civil engineering for scientific research, and instruments and equipment for scientific research (Figure 1-8). The proportions of these two items in 2016 were respectively 44.52% and 54.82%.

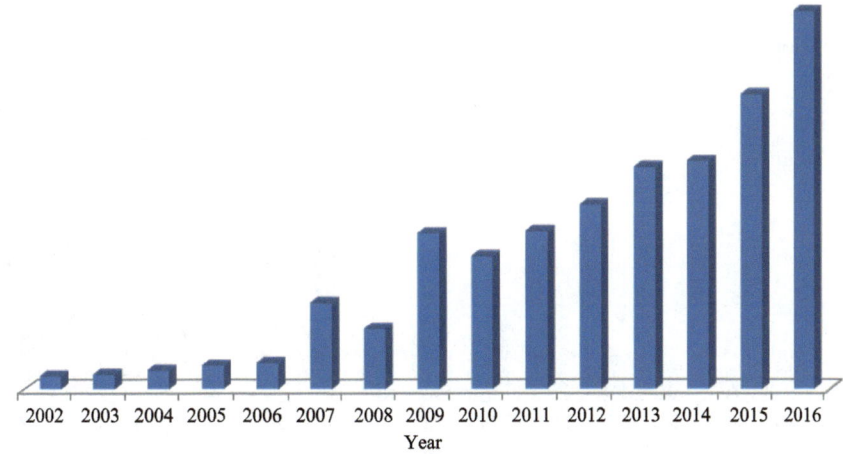

Figure 1-7 Actual Completion of Capital Construction Investment of Marine Scientific Research Institutions from 2002 to 2016 (Thousand Yuan)

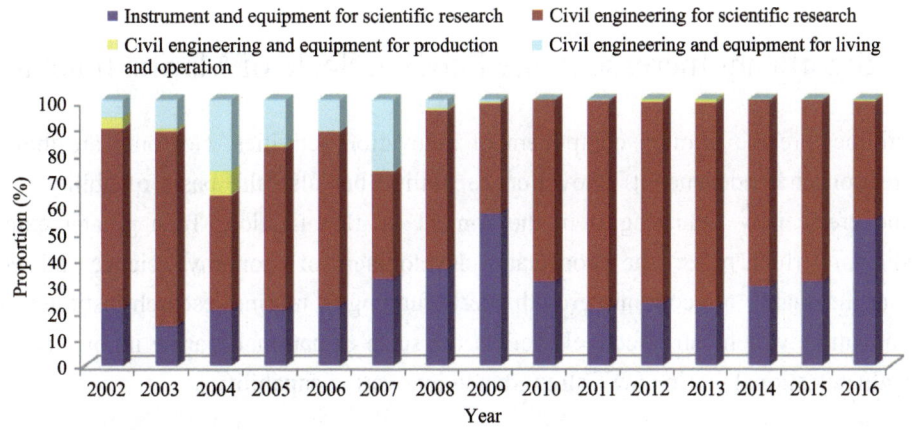

Figure 1-8 Actual Completion Composition of Capital Construction Investment of Marine Scientific Research Institutions from 2002 to 2016

1.2.3 Year-on-year increase in fixed assets, and scientific instruments and equipments

Fixed assets refer to the facilities and equipments which can be used for a long time, the value of which can be consumed, but the original physical form can be maintained, such as houses and buildings. Properties and materials as fixed assets should meet two requirements at the same time: a durability period of more than one year and a unit value higher than the specified standard. From 2002 to 2016, the original price of fixed assets of marine scientific research institutions in China has continued to grow (Figure 1-9), with an average annual growth rate of 22.04%. Scientific instruments and equipment in the original price of fixed assets refer to the instruments and equipment for scientific research directly used by personnel engaged in S&T activities excluding all kinds of power equipment, machinery equipment, auxiliary equipment which support capital construction, and general transportation equipment (other than means of transportation for scientific investigation), and instruments and equipment designated for production. From 2002 to 2016, the original price of the scientific instruments and equipment in fixed assets of China's marine scientific research institutions also has maintained a growth trend (Figure 1-9), with an average annual growth rate of 24.88%.

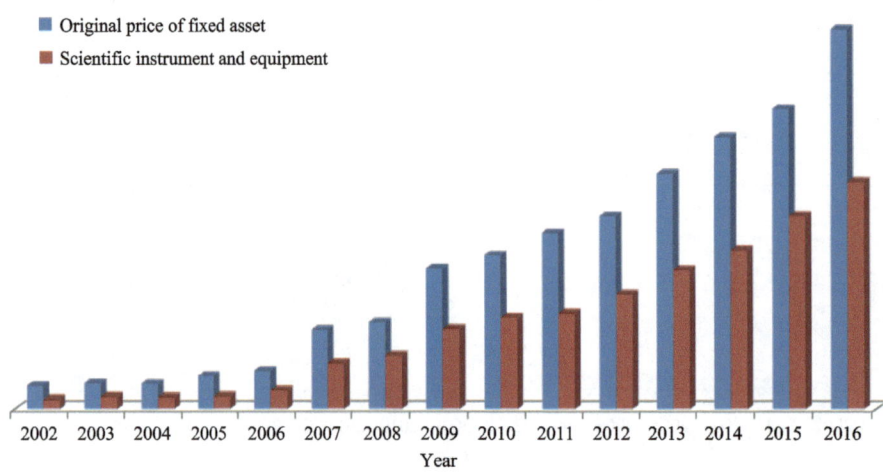

Figure 1-9　Original Price of Fixed Assets (Thousand Yuan) and the Scientific Instruments and Equipment (Thousand Yuan) in Original Price of Fixed Assets of Marine Scientific Research Institutions from 2002 to 2016

1.3　Significant Increase in the Funding Scale of Marine Innovation

R&D activities are the primary component of innovation activities. Not only are they a source of knowledge creation and independent innovation capability, but also the basis of ability to absorb new knowledge and create new technologies in the context of globalization. To a greater extent, they are important indicators which reflect the coordinated development of economy, science and technology, as well as measure the quality of economic growth. R&D funding of marine research institutes are important for marine innovation, which can effectively reflect the scale of national marine innovation activities and objectively evaluate national marine S&T strength and innovation capability.

1.3.1　Steady increase of R&D funding

From 2002 to 2016, the R&D funding in China's marine scientific research institutions has maintained a growth trend with annual growth rate reaching 23.72% (Figure 1-10). The year 2007 witnessed the most rapid growth in this indicator, with an annual growth rate reaching 145.18%.

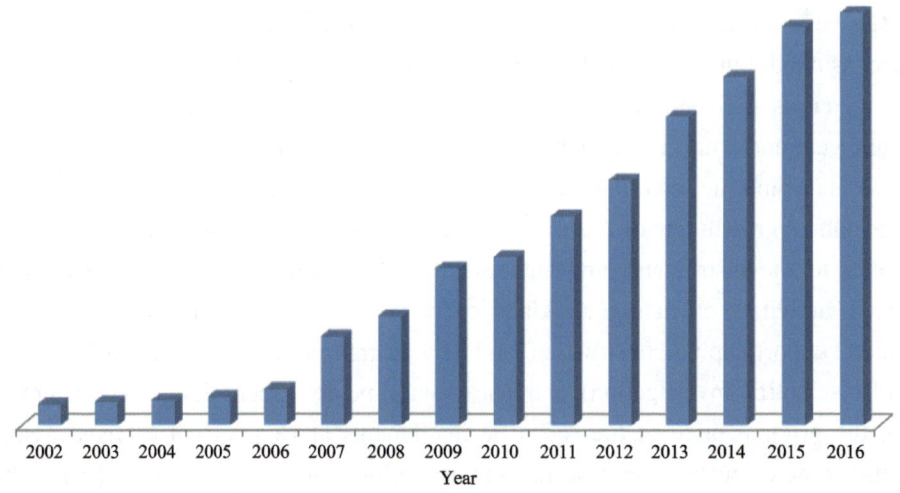

Figure 1-10　R&D funding (Thousand Yuan) from 2002 to 2016

The proportion of R&D funds in the national Gross Ocean Product (GOP) generally acts as the indicator revealing the investment strength of national marine scientific research funds, reflecting the input strength of national marine innovation funds. From 2002 to 2016, this indicator has conveyed an overall growth trend, with an average annual growth rate of 8.54%. This indicator decreased slightly in 2016 compared to 2015 (Figure 1-11).

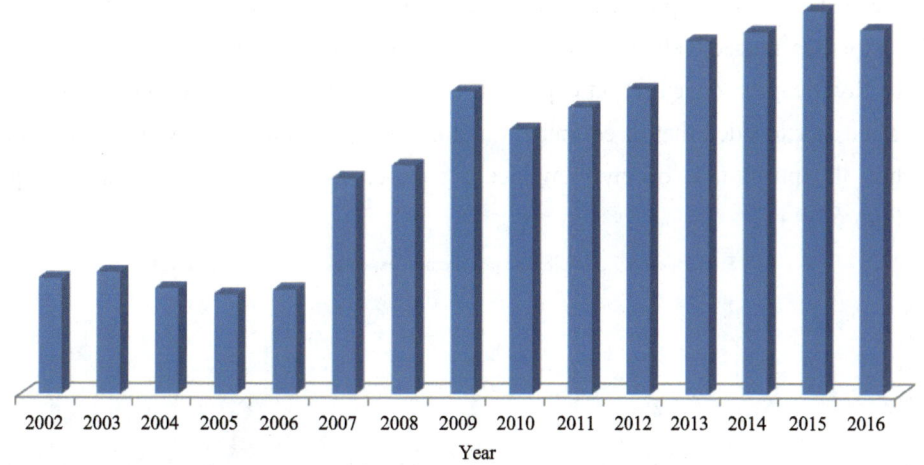

Figure 1-11　Proportion of R&D Funding in National GOP from 2002 to 2016

1.3.2　Steady growth of the internal spending of R&D funding

The internal spending of R&D funding refers to all expenditures incurred in the current year for carrying out R&D activities by the institutions, including R&D regular expenses and R&D capital construction expenses. From 2002 to 2016, the proportion of R&D capital construction expenses in the internal spending of R&D funding has shown a fluctuating trend. The proportion increased from 8.71% in 2002 to 13.98% in 2016, reflecting the greater emphasis placed on capital construction investment in China (Figure 1-12).

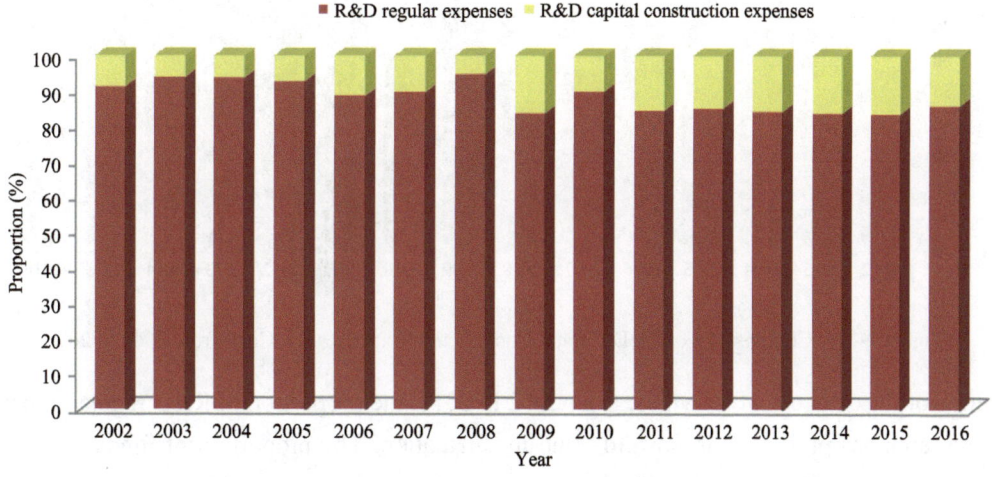

Figure 1-12　Composition of Internal Spending of R&D Funds from 2002 to 2016

In terms of the category of expenses, R&D regular expenses include staff expenses (including salary), equipment procurement expenses and other daily expenses (including operating expenses and overheads), and R&D capital construction expenses include instruments and equipment expenses and civil engineering expenses. From 2002 to 2016, other daily expenses accounted for over 50% in R&D regular expenses, and the proportion of staff expenses and equipment procurement expenses witnessed a minor decline (Figure 1-13). In 2016, staff expenses and equipment procurement expenses accounted for 30.15% and 12.08% of the R&D regular expenses respectively and other daily expenses took up 57.77%. From 2002 to 2016, the composition of R&D capital construction expenditures showed a fluctuating trend, the proportion of civil engineering expenses exceeded that of equipment and instruments in all other years except the years 2007 and 2009 when the proportion of civil engineering expenses was less than that of equipment and instruments (Figure 1-14).

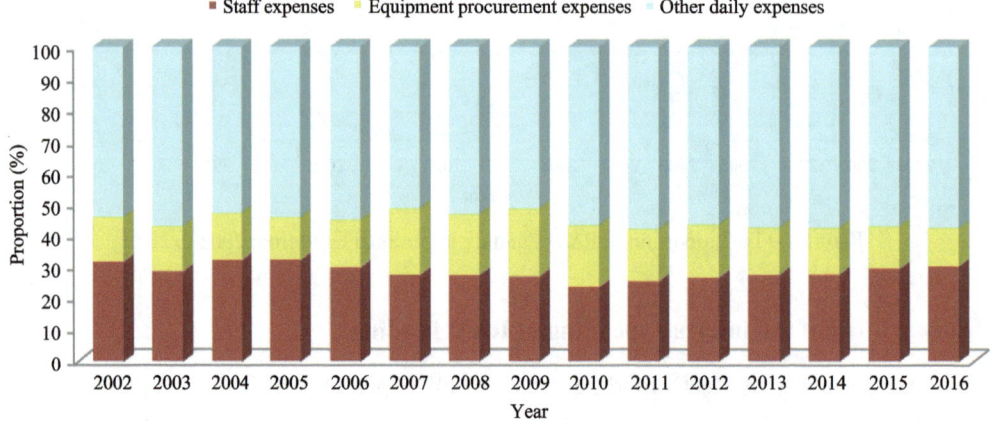

Figure 1-13   Composition of Expenditures of R&D Regular Expenses (by Cost) from 2002 to 2016

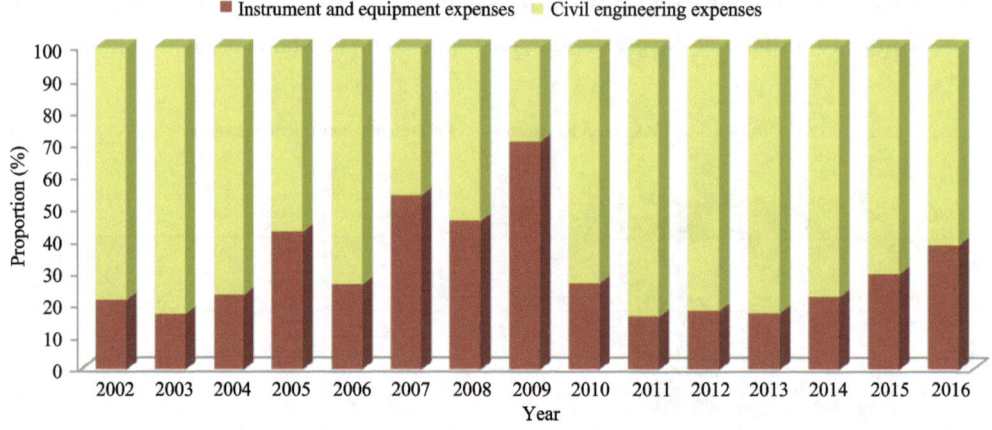

Figure 1-14   Composition of R&D Capital Construction Expenses (by Cost) from 2002 to 2016

In terms of the category of activities, from 2002 to 2016, the proportion of R&D regular expenses used for basic research has presented an upward trend in fluctuation. The proportion of funds used for applied research fell from 48.73% in 2002 to 37.73% in 2016, and the proportion of funds used for experimental development dropped from 33.05% in 2002 to 31.28% in 2016 (Figure 1-15).

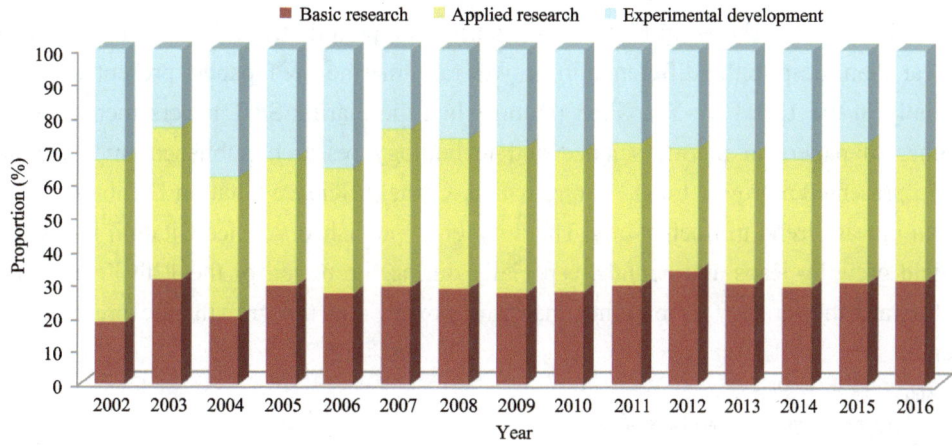

Figure 1-15　Composition of R&D Regular Expenses (by Category of Activity) from 2002 to 2016

In terms of the sources of funds, from 2002 to 2016, the internal spending of R&D funds mainly derived from governmental funds and corporate funds. In addition, the proportion of governmental funds swayed downward, while that of corporate funds increased. In 2016, the proportions of governmental funds and corporate funds were 80.69% and 9.11% respectively (Figure 1-16).

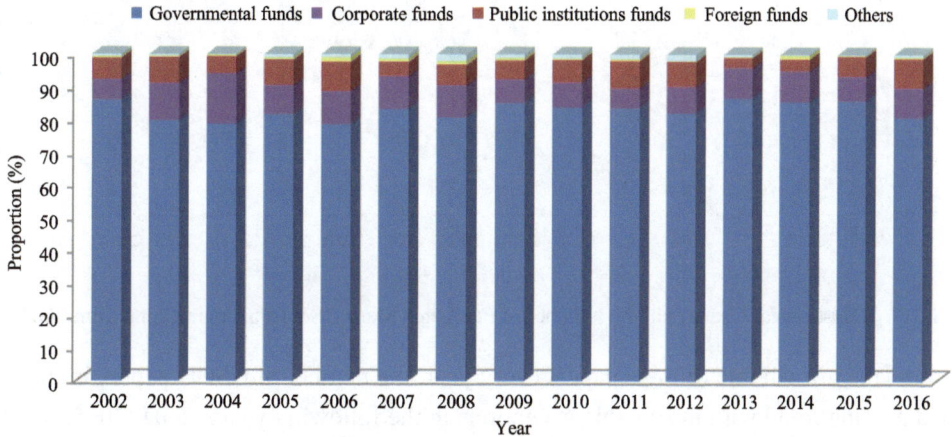

Figure 1-16　Composition of Internal Spending of R&D Funds (by Source) from 2002 to 2016

1.4　Continuous Growth of Marine Innovation Output Achievements

Knowledge innovation is a core element of competitiveness among nations. Innovation output means intermediate achievements of various types produced by scientific research and technological innovation activities. The quality and quantity of academic papers and writings reflect the innovation capability of marine science and technology, while patent applications and grants more directly reflect the degree of marine innovation activities and level of technological innovation. Relatively higher ability in marine knowledge diffusion and application is one of the common features of an innovative maritime power.

1.4.1　Positive growth trend in the total number of marine S&T papers

From 2001 to 2016, the total number of S&T papers in the marine field has maintained an overall growth trend. The number of papers published in 2016 was 6.06 times that of 2001, with an average annual growth rate of 12.77%. As shown in Figure 1-17, during the periods from the 11th Five-Year Plan to the 12th Five-Year Plan, despite the difference in growth rate, marine S&T papers presented a linear growth trend, especially in the 12th Five-Year Plan period when the marine S&T papers increased rapidly. The annual number of papers in marine science and technology research published in China and foreign countries also presented a growth trend, among which, Chinese Science Citation Database (CSCD) papers maintained an upward trend in fluctuations. The number of published Science Citation Index (SCI) papers in marine field grew by leaps and bounds, especially during the period of the 12th Five-Year Plan since China put forward the strategy of building maritime power, and the growth rate presented an obvious increase trend.

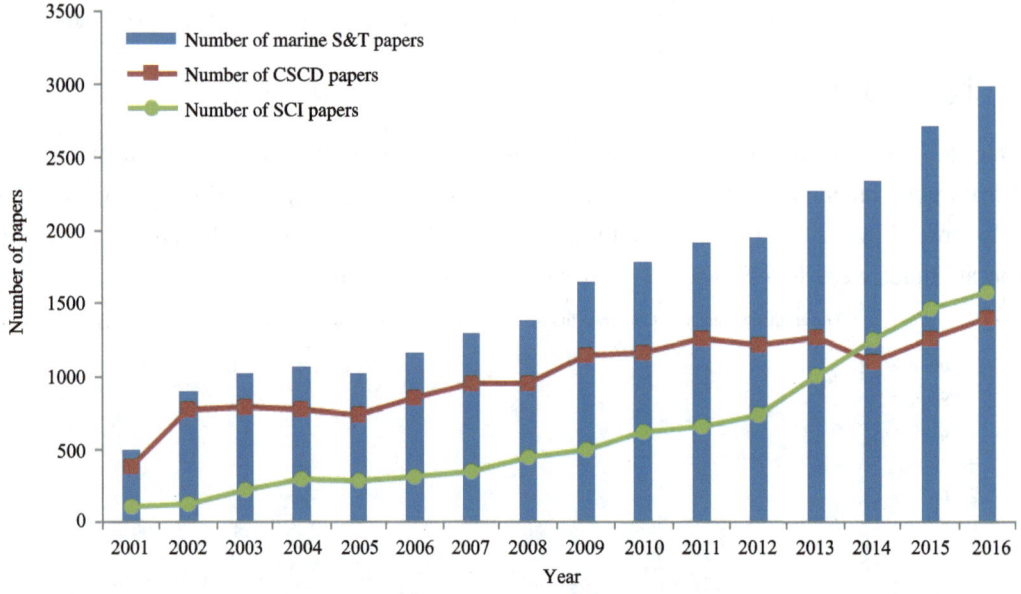

Figure 1-17　Year-on-year Changes in the Number of Published Marine S&T Papers in China from 2001 to 2016

In terms of the growth rate of S&T papers every year, the number of CSCD papers in the marine field has presented a rising trend with the notable exceptions in the following years: 2004, 2005, 2008, 2012 and 2014. The annual growth rates of 2006 and 2009 were both above 15%. Except 2005, the number of SCI papers in marine field published every year has shown a positive increasing trend. The years when the growth rates of 25% and above were 2003, 2004, 2008, 2013 and 2014 (Table 1-1).

Table 1-1　Number of Published Marine S&T Papers in China and Analysis of Annual Growth Rate from 2001 to 2016

Year	Number of CSCD papers	Number of SCI papers	Number of marine S&T papers	Annual growth rate (%)	
				Number of CSCD papers	Number of SCI papers
2001	384	108	492		
2002	767	126	893	100	17
2003	791	224	1015	3	78
2004	772	294	1066	−2	31
2005	737	279	1016	−5	−5
2006	853	308	1161	16	10
2007	948	346	1294	11	12
2008	943	442	1385	−1	28

续表

Year	Number of CSCD papers	Number of SCI papers	Number of marine S&T papers	Annual growth rate (%)	
				Number of CSCD papers	Number of SCI papers
2009	1144	499	1643	21	13
2010	1161	619	1780	1	24
2011	1257	655	1912	8	6
2012	1217	736	1953	−3	12
2013	1265	1000	2265	4	36
2014	1097	1246	2343	−13	25
2015	1254	1461	2715	14	17
2016	1403	1580	2983	12	8

1.4.2 Increase in both quantity and quality of SCI papers in oceanography in China

From 2001 to 2016, there were 9923 SCI papers published in oceanography in China, with annual published papers presenting an obvious increasing trend, especially after the year 2012 (Figure 1-18). The number of published papers in 2016 was 14.63 times that of 2001. From 2001 to 2016, the number of SCI papers conveyed an obvious increasing trend, with the year 2013 as the turning point in the increase of SCI papers. The SCI papers increased in fluctuations from 2006 to 2010 and the periods from 2011 to 2016 witnessed an initial increase followed by a decreasing trend in the number of SCI papers. As shown in Figure 1-19, from 2001 to 2016, the number of SCI papers in international oceanography both increased and decreased. The number of publications in China has presented a continual growth trend, particularly after entering the 12th Five-Year Plan when the increase was fast. As shown in Figure 1-20, from 2001 to 2016, the number of SCI papers with China as the first author presented an increasing trend, among which, the period from 2012 to 2015 presented an obvious linear growth.

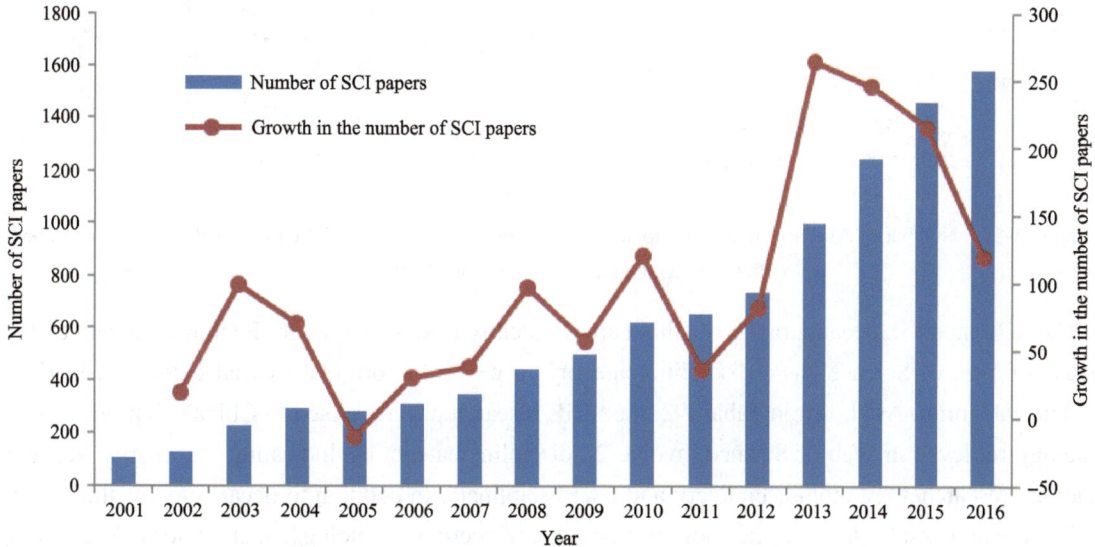

Figure 1-18　Number of SCI Papers in Oceanography Published Annually in China and Incremental Change from 2001 to 2016

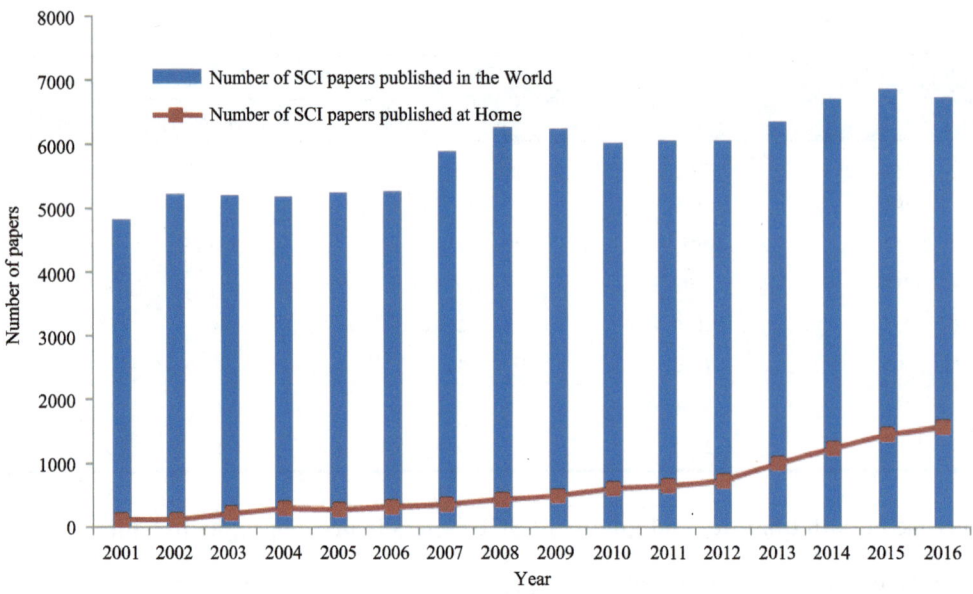

Figure 1-19　Number of SCI Papers in Oceanography Published at Home and the World from 2001 to 2016

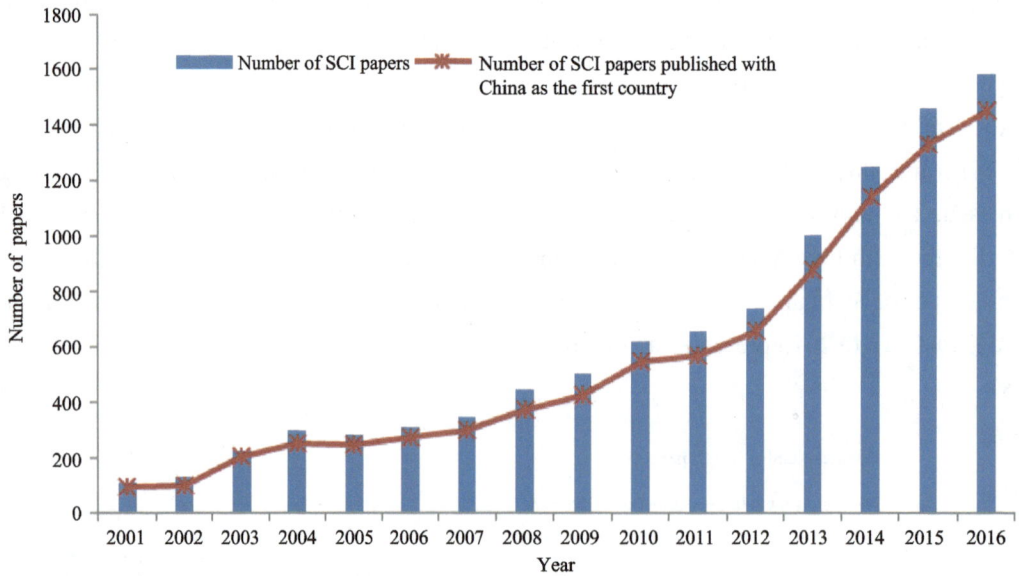

Figure 1-20　Number of All Published SCI Papers in Oceanography of China and Number of Published SCI Papers with China as the First Country from 2001 to 2016

The SCI papers in oceanography of China are frequently cross-disciplined. Each record included in the database of Web of Science has a discipline category to which the original journal belongs, covering 252 discipline categories. As shown in Table 1-2, the SCIE research papers related to China's marine science and technology retrieved in Web of Science involve 22 discipline categories, indicating that marine science and technology researches are multidisciplinary and interdisciplinary. In addition to oceanography, the disciplines that the research results involve the most are ocean engineering, limnology, meteorology & atmospheric science, science of water resources, marine & freshwater biology, ecology, geochemistry & geophysics, geological engineering, mining and mineral processing, fisheries, and other disciplines related to marine science and technology and interdisciplinary fields.

Table 1-2　Disciplinary Distribution of SCIE Papers in China's Marine Science and Technology from 2001 to 2016

Sequence number	WOS discipline classification	Number of papers
1	Oceanography	10 389
2	Ocean Engineering	4 634
3	Civil Engineering	2 508
4	Limnology	1 350
5	Meteorology & Atmospheric Science	1 220
6	Science of Water Resources	1 213
7	Geosciences, Multidisciplinary	1 206
8	Marine & Freshwater Biology	1 094
9	Mechanical Engineering	1 054
10	Engineering, Multidisciplinary	750
11	Ecology	233
12	Geochemistry & Geophysics	200
13	Geological Engineering	157
14	Mining and Mineral Processing	157
15	Fisheries	143
16	Chemistry, Multidisciplinary	141
17	Electronic & Electrical Engineering	115
18	Remote Sensing	62
19	Environmental Science	45
20	Mechanics	45
21	Paleontology	43
22	Energy & Fuels	3

The journal distribution of SCI papers in marine field is shown in Table 1-3, which makes statistical analysis of the top 20 journals which have published SCI papers in the field of marine science and technology of China. The journals which published more than 1000 papers are *Acta Oceanologica Sinica*, *Chinese Journal of Oceanology and Limnology*, followed by *China Ocean Engineering*, *Ocean Engineering*, *Journal of Ocean University of China* and *Journal of Geophysical Research-Oceans* which published more than 500 papers.

Table 1-3　Major Journals Publishing the SCI Papers of China's Marine Science and Technology from 2001 to 2016

Sequence number	Name of journals	Number of published papers	Sequence number	Name of journals	Number of published papers
1	Acta Oceanologica Sinica	1479	6	Journal of Geophysical Research-Oceans	518
2	Chinese Journal of Oceanology and Limnology	1173	7	Estuarine Coastal and Shelf Science	320
3	China Ocean Engineering	874	8	Continental Shelf Research	294
4	Ocean Engineering	709	9	Journal of Navigation	186
5	Journal of Ocean University of China	554	10	Applied Ocean Research	185

Continued

Sequence number	Name of journals	Number of published papers	Sequence number	Name of journals	Number of published papers
11	Marine Ecology Progress Series	159	16	Marine Geology	125
12	Marine Georesources & Geotechnology	154	17	Deep-Sea Research Part II-Topical Studies in Oceanography	123
13	Journal of Marine Systems	138	18	Journal of Oceanography	122
14	Terrestrial Atmospheric and Oceanic Sciences	138	19	Ocean & Coastal Management	117
15	Journal of Atmospheric and Oceanic Technology	133	20	Marine Chemistry	112

Table 1-4 indicates the top 20 organizations which have published SCI papers in the field of marine science and technology. The Chinese Academy of Sciences ranked the 1st, the number of its published papers was more than 1.5 times that of Ocean University of China which ranked the 2nd. The organizations which have published more than 1000 papers include original State Oceanic Administration[①] ranking the 3rd, followed by Dalian University of Technology, Xiamen University, Shanghai Jiao Tong University, East China Normal University, Zhejiang University, Hohai University and Chinese Academy of Fishery Sciences.

Table 1-4   Major Institutions with Published SCI Papers in the Field of Marine Science and Technology and Number of Published Papers from 2001 to 2016

Sequence number	Name of institutions	Number of papers
1	Chinese Acad. Sci.	2709
2	Ocean Univ. China	1783
3	Original State Ocean. Adm.	1258
4	Dalian Univ. Technol.	588
5	Xiamen Univ.	518
6	Shanghai Jiao Tong Univ.	457
7	East China Normal Univ.	318
8	Zhejiang Univ.	316
9	Hohai Univ.	287
10	Chinese Acad. Fishery Sci.	248
11	Hong Kong Univ. Sci. & Technol.	219
12	Tianjin Univ.	218
13	Tongji Univ.	200
14	Shanghai Ocean Univ.	199
15	Harbin Engn. Univ.	197
16	Nanjing Univ.	165
17	Woods Hole Oceanog. Inst.	124
18	Sun Yat-sen Univ.	114
19	Univ. Hong Kong	137
20	Nanjing Univ. Informat. Sci. & Technol.	130

① In March 2018, the State Council launched an institutional reform to set up the Ministry of Natural Resources, which incorporated the responsibilities of the State Oceanic Administration into it. But the State Oceanic Administration still retains its brand to the outside world

1.4.3 Significant increase in the categories of marine S&T publications

From 2002 to 2016, the categories of marine S&T publications by China's marine scientific research institutions has presented a significant growth trend (Figure 1-21), with an average annual growth rate of 13.97%, among which, the category of marine S&T publications has been at a steady growth stage from 2002 to 2005 with an average annual growth rate of 17.27%; the 2006-2007 and 2008-2009 periods witnessed rapid growth in the categories of marine S&T publications, with growth rates of 104.41% and 64.47% respectively; the average annual growth rate of the categories of marine S&T publications from 2010 to 2016 was 10.64%.

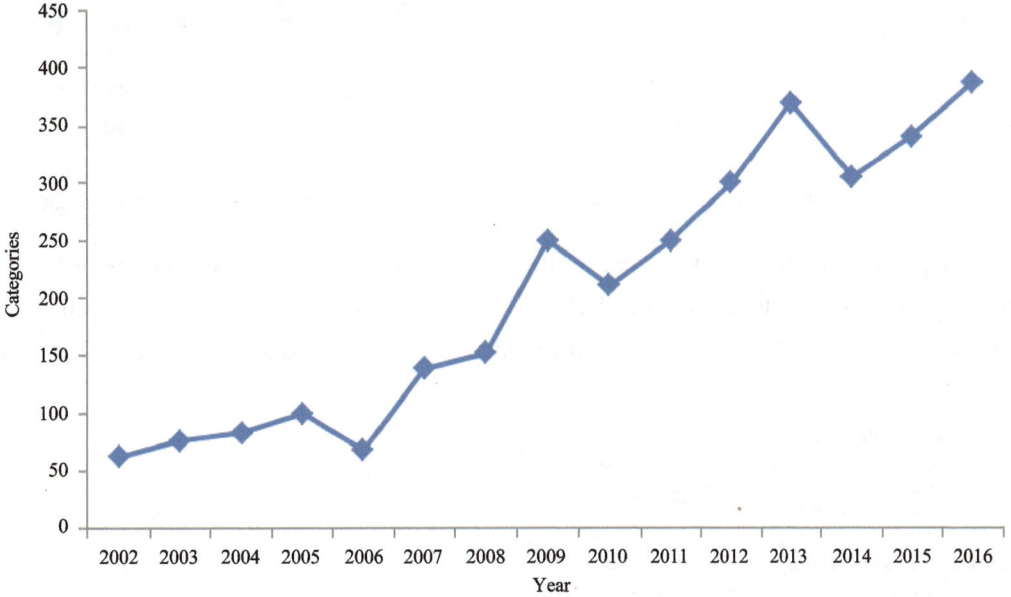

Figure 1-21　Marine S&T Publications Categories of China from 2002 to 2016

1.4.4 First increase and then descend in marine engineering technology (EI) papers

Engineering Index (EI) is a comprehensive information retrieval tool for the literature in the field of engineering technology founded by the American Association of Engineering Societies, which is recognized as a very important retrieval tool by the engineering industry. It includes 20 million pieces of data, covering 76 countries and more than 190 engineering disciplines, more than 3 600 journals and periodicals, more than 80 book serials, more than 80 000 conference proceedings, more than 80 000 dissertations, and hundreds of trade magazines. This report has done statistical analysis of papers related to the marine field in the EI database to understand the research and development trends of the world and China in the marine field. From 2001 to 2016, there were 275 611 related publications in the field of global oceanography and 55 876 publications were related to China. As is shown in Figure 1-22, the number of EI papers in China's marine field and its global share rose steadily. China's growth rate in the past 15 years has far exceeded the global growth rate. In 2012, the number of China's marine EI papers was 31 times higher than in 2001, and its global share increased from 4.74% in 2001 to more than 25% in 2016.

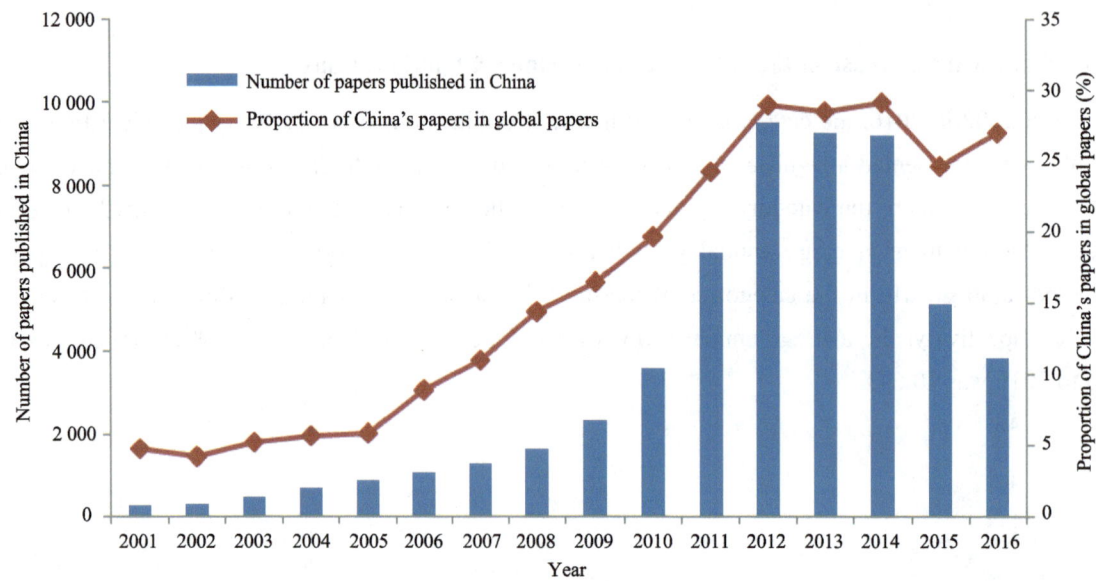

Figure 1-22   Number of EI Papers in China's Marine Field and Its Proportion in Global EI Papers from 2001 to 2016

The disciplinary distribution of papers in the marine field of China mirrors the world's publication rate, but the papers in the naval vessels field in China account for a larger proportion. Disciplines related to chemistry (such as general chemical products, organic compound, etc.) are widely distributed (Figure 1-23).

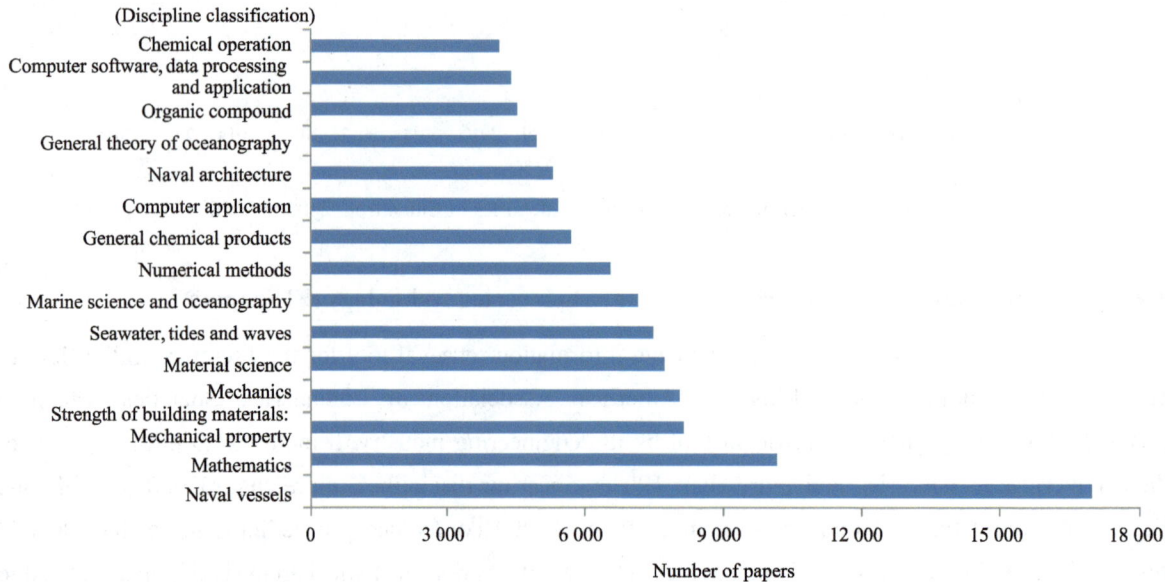

Figure 1-23   Disciplinary Distribution of EI Papers Output in China's Marine Field

The 15 institutions with most EI papers in the marine field in China are shown in Figure 1-24. There are many marine-related research institutions in China, where the number of S&T papers of the Chinese Academy of Sciences in marine field accounts for 8.32% of the country.

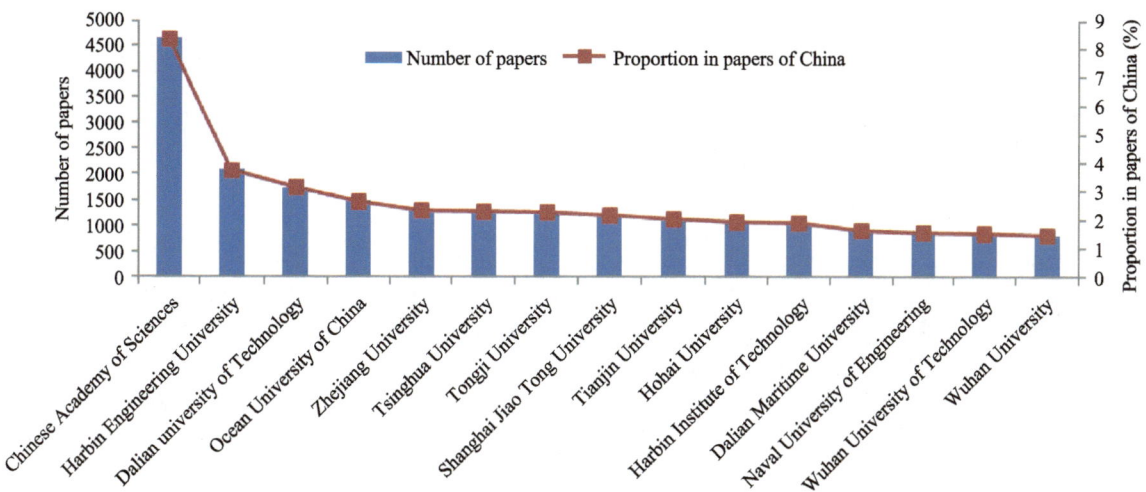

Figure 1-24　Number of EI Papers in the Marine Field of China Published by the Top 15 Institutions and Their Proportions in National Share

1.4.5　Vigorous growth in the number of patent applications in marine field

From 2001 to 2016, the number of patent applications in marine field has been on the rise every year with an average annual growth rate at 23.90%. The increase has been significant since 2006. As shown in Figure 1-25 and Figure 1-26, the number of annual patent applications has maintained above 4000 patents from 2012 to 2016. China's marine patents ushered in a period of a rapid development period in 2006, increasing annually by more than 700 from 2010 to 2016. Due to the hysteretic nature of patent data, the data in the past three years are only for reference, but we can still see that the development of marine technology in China is in a period of high-speed growth.

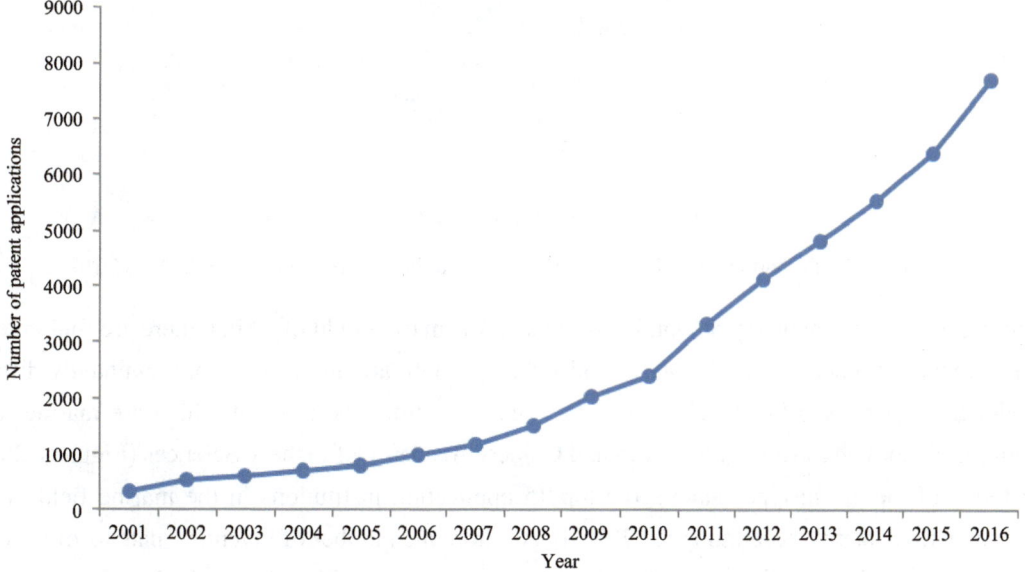

Figure 1-25　Number of Patent Applications in the Marine Field of China from 2001 to 2016

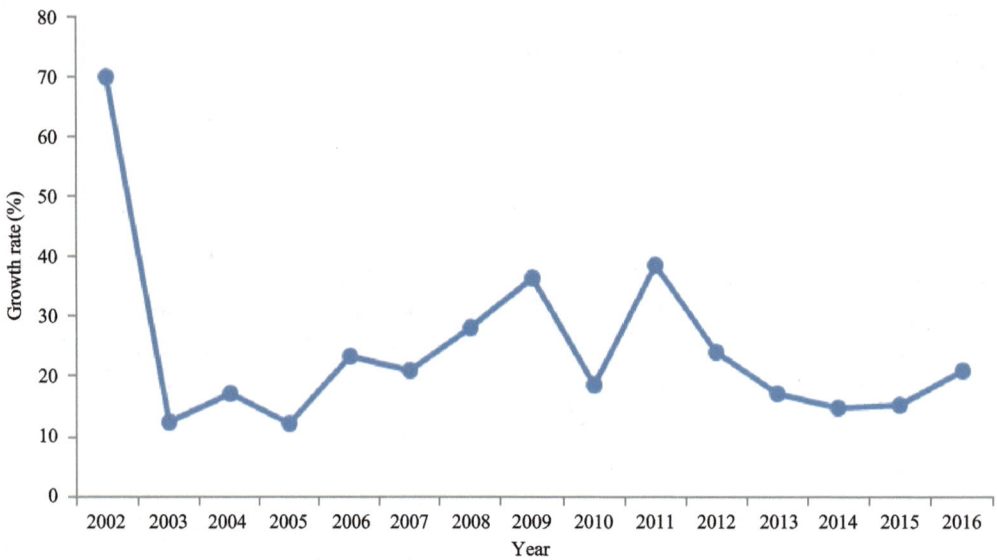

Figure 1-26   Annual Growth Rate Change in the Number of Patent Applications in the Marine Field of China from 2002 to 2016

Invention patents account for 62% of patents in the marine field of China (Figure 1-27), indicating that currently most of the marine patents focus on technology and R&D, having great potential for innovation.

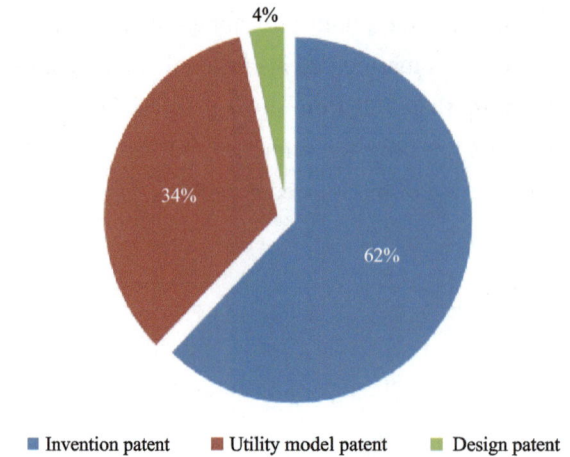

Figure 1-27   Proportion of Patent Categories in the Marine Field of China from 2001 to 2016

Among the top 15 patent application institutions in the marine field of China, there are four enterprises with all of them contained within the marine oil industry. Eight are universities, predominantly distributed in Shandong, Zhejiang and Shanghai. Three are research institutions, most of which are marine research institutions of Chinese Academy of Sciences and Chinese Academy of Fishery Sciences (Figure 1-28).

In terms of the patent types among the top 15 application institutions in the marine field of China, most are invention patents, accounting for 70%. In addition, the number of invention patents of universities and scientific research institutions is higher than that of enterprises. There are only 2 design patents, with one patented by China National Offshore Oil Corporation and the other by Harbin Engineering University (Figure 1-29).

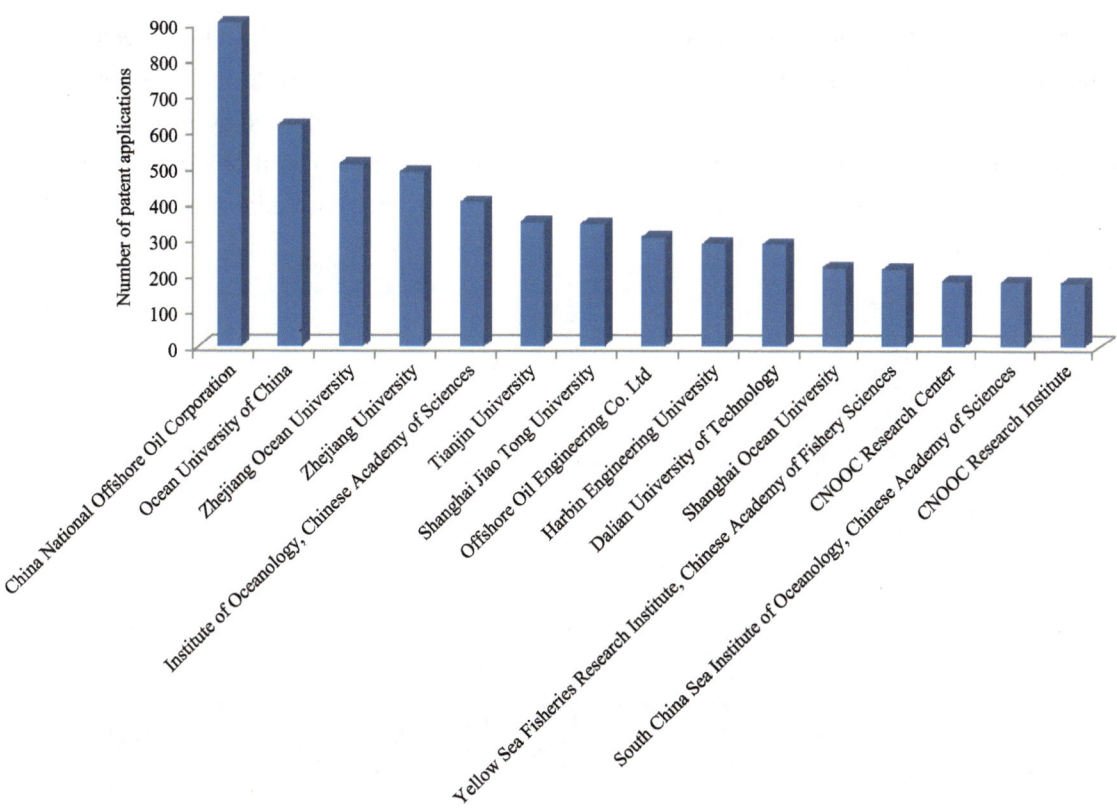

Figure 1-28　Major Patent Application Institutions in the Marine Field of China from 2001 to 2016

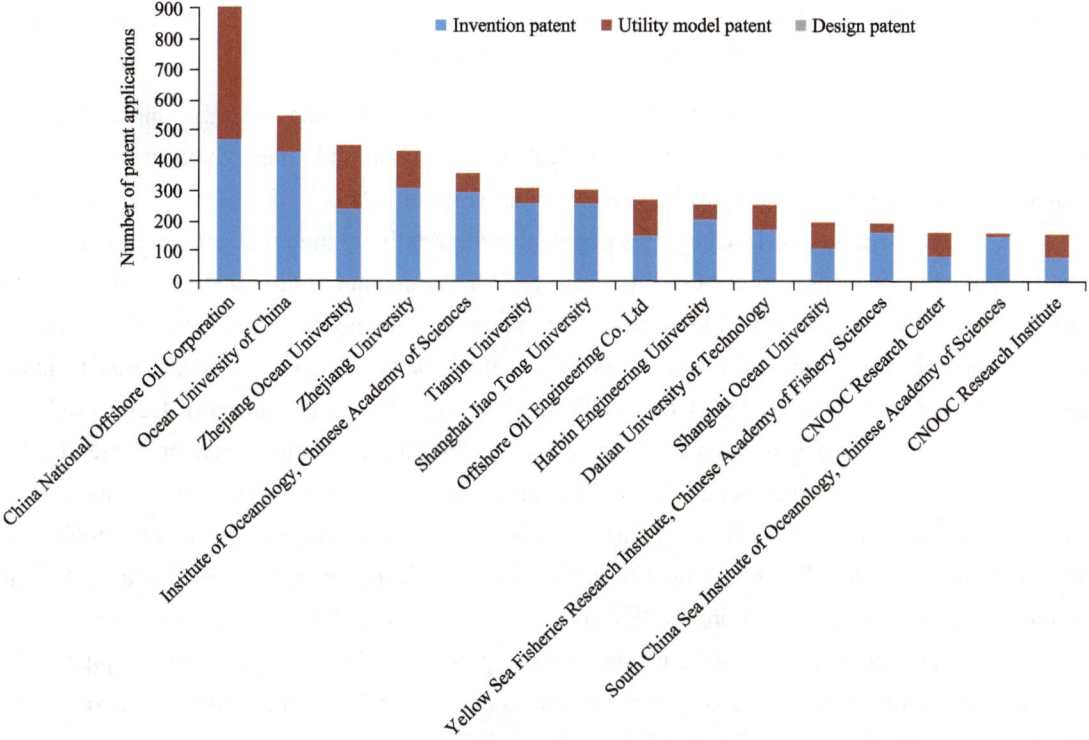

Figure 1-29　Categories of Major Patent Application Institutions in the Marine Field of China from 2001 to 2016

The design patent shown in the figure is not obvious due to low number of applications

From 2001 to 2016, among the major patent application provinces (municipalities directly under the central government) in the marine field of China, Shandong ranked first due to its comparatively larger number of marine-related scientific research institutions and universities, followed by Jiangsu and Zhejiang which ranked the second and third. Beijing ranked fourth. Among other coastal provinces (municipalities directly under the central government), the number of patent applications of Fujian Province was relatively small (Figure 1-30).

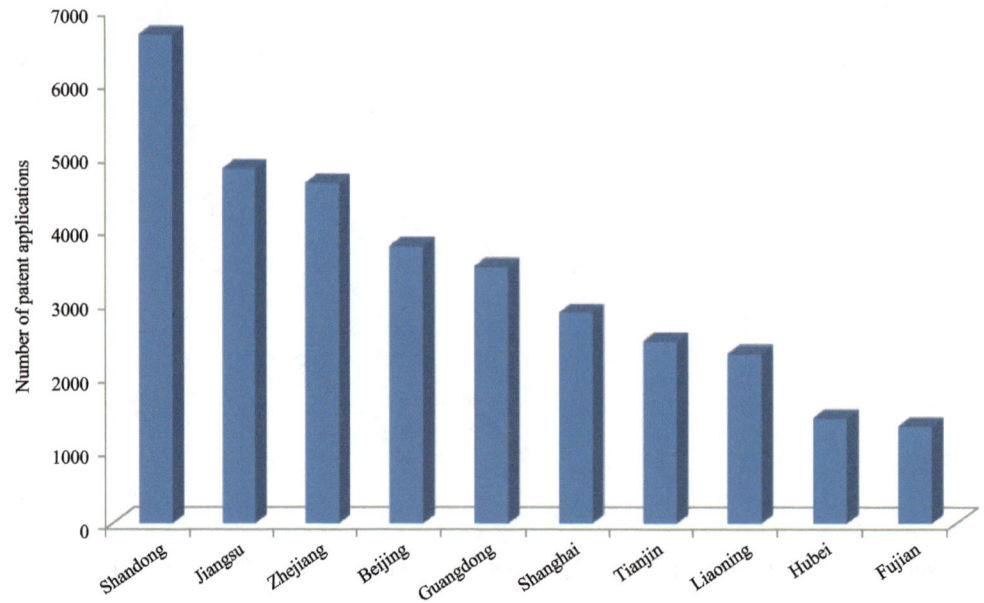

Figure 1-30　Major Patent Application Provinces and Municipalities Directly under the Central Government in the Marine Field of China from 2001 to 2016

In terms of major applicantion cities, the number in Qingdao was considerably higher than that of other cities. The number of patent applications in Dalian, Hangzhou and Guangzhou was quite high, and Wuhan and Nanjing were outstanding as non-coastal cities (Figure 1-31).

From 2001 to 2016, the top 15 categories of patents with relatively higher occurrence frequency in the marine field of China were as follows: C02F (sewage and sludge contamination treatment), A01K (fisheries management; fish farming), B63B (ship or other waterborne vessels; marine equipment), G01N (testing or analyzing materials by means of chemical or physical property of the materials), F03B (hydraulic machinery or hydraulic motors), E02B (hydraulic engineering), B01D (separation), E21B (soil or rock drilling), A61K (medical preparations and formulations), E02D (foundation; excavation; embankment; underground or underwater structure), A23L (food, foodstuff or non-alcoholic beverages not included in the A21D or subcategories from A23B to A23J), C12N (micro-organisms or enzymes), C09D (coating composition, such as colored paint, varnish or natural lacquer; filling slurry; chemical coating or printing ink remover; printing ink; correction fluid; wood stain; slurry or solids used for staining or printing; application of raw materials for this purpose), F16L (pipe; pipe joint or pipe fitting; pipe, cable or protective pipe support; general adiabatic method), H01B (cable; conductor; insulator; choice of conductive, insulating or dielectric materials) (Figure 1-32).

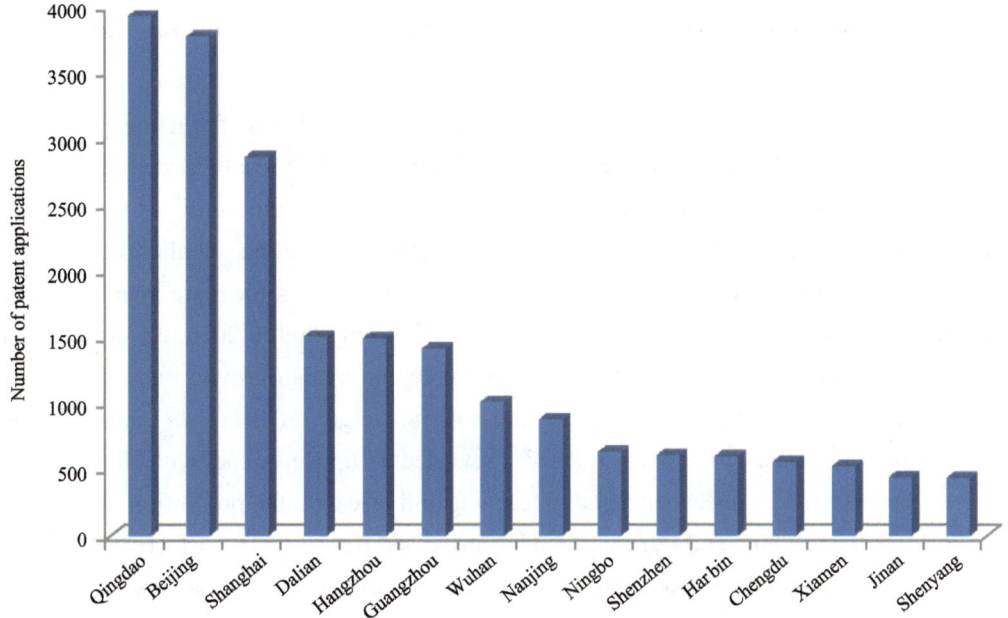

Figure 1-31　Major Patent Application Cities in the Marine Field of China from 2001 to 2016

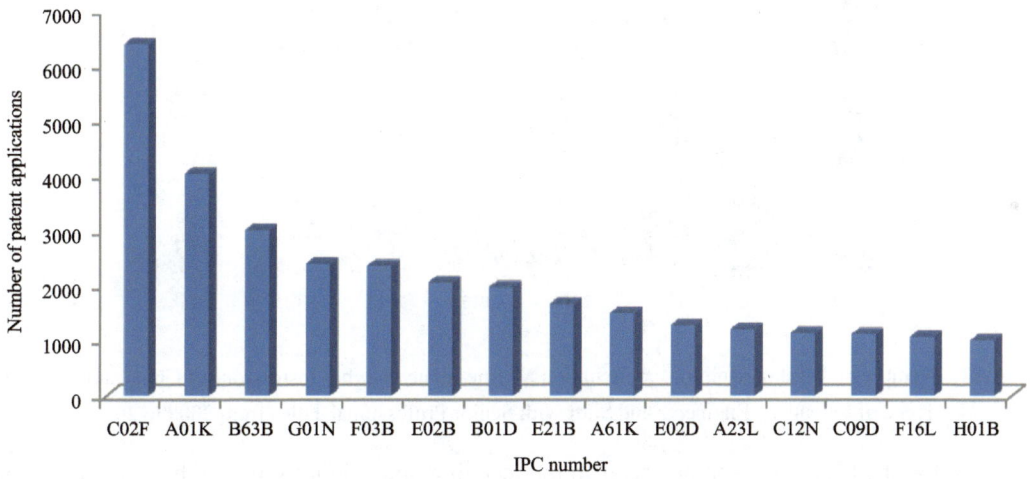

Figure 1-32　Number of Major International Patent Classification in the Marine Field of China from 2001 to 2016

1.5　Steady Improvement in Marine Innovation of Higher Institutions

Higher institutions play a decisive role in national innovation development. In recent years, both the input of marine innovation resources and the output of marine innovation achievements at higher institutions of China have gradually been boosted and marine innovation has developed healthily. It should be noted that the data used in this section are extracted on the basis of marine-related higher institutions and marine-related disciplines and then obtained by weighted summation according to their marine-related coefficient of proportionality (See Appendix 8 for the list of marine-related higher institutions and marine-related coefficient of proportionality and Appendix 9 for marine-related disciplines and marine-related coefficient of proportionality).

1.5.1 Gradual optimization in the structure of human resources for marine innovation in higher institutions

The teaching and research staff at higher institutions refer to the staff on the payroll of higher institutions who are engaged in teaching, R&D, and application of R&D results at the junior college level and above, plus the technology service staff and the personnel who provide services for the above work during the statistical year, including foreign experts, experts and visiting scholars outside the higher education system who have been engaged in scientific research activities for more than one month in the aggregate within the statistical year. As shown in Figure 1-33, from 2009 to 2016, the number of teaching and research staff at higher institutions in China has presented an overall growth trend. Since 2015, there has been a slight decline, but the overall fluctuation was not large. Of them, the number of scientists, engineers and staff with senior professional titles also revealed an upward trend, with slight fluctuations in the number of scientists and engineers engaged in teaching and research; the proportion of staff with senior professional titles in the teaching and research field has increased from 37.50% to 43.10%.

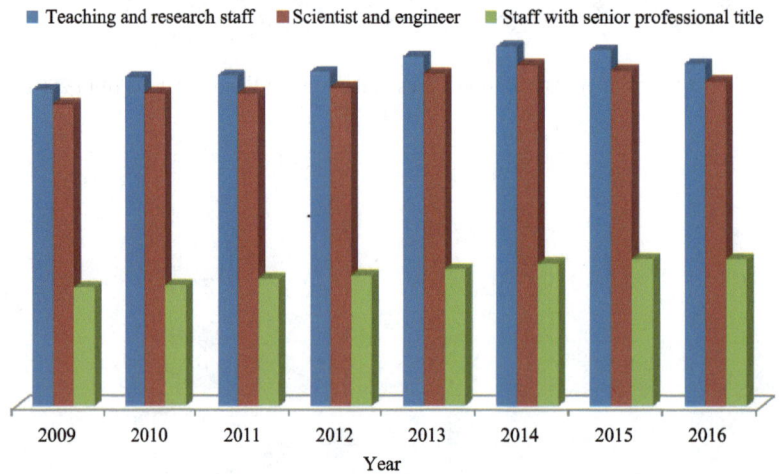

Figure 1-33 Number of Teaching and Research Staff at Marine-related Higher Institutions of China and the Variation Trend of Scientists, Engineers and Staff with Senior Professional Titles from 2009 to 2016

R&D staff at the higher institutions refer to the teaching and scientific research personnel whose time spent in R&D accounts for more than 10% of the total time spent in their own teaching and scientific research within the statistical year. From 2009 to 2016, the number of R&D staff at the marine-related colleges and universities in China were basically stable (Figure 1-34). Of them, the number of scientists, engineers and staff with senior professional titles also presented an upward trend, and the proportion of scientists and engineers among the R&D staff dropped slightly, decreasing from 95.99% to 95.47%. There have been slight fluctuations in the proportion of staff with senior professional titles among the R&D staff.

1.5.2 Gradual increase of the input in marine innovation of higher institutions

From 2009 to 2016, China's marine-related colleges and universities have made increasing investment in science and technology funds, with an average annual growth rate of 11.38%. The governmental funds have shown a growing trend, with an average annual growth rate reaching 12.12%. The internal spending of marine-related higher institutions in China in the current year has shown a significant growth (Figure 1-35), and the internal spending in 2016 was 66.98 times that of 2009.

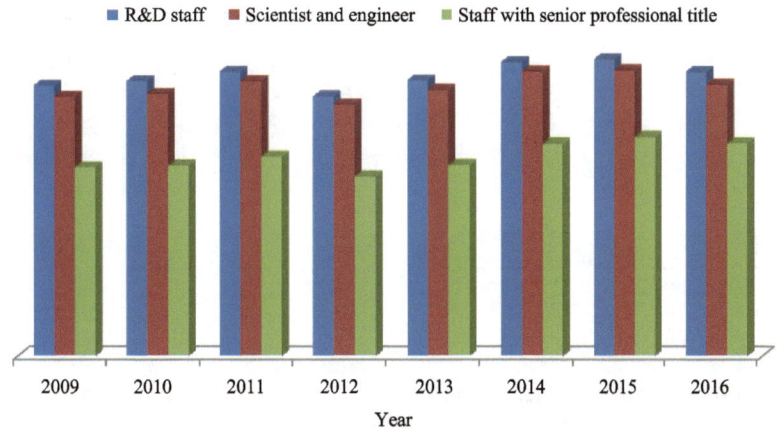

Figure 1-34 Number of R&D Staff at Marine-related Higher Institutions of China and the Variation Trend of Scientists, Engineers and Staff with Senior Professional Titles from 2009 to 2016

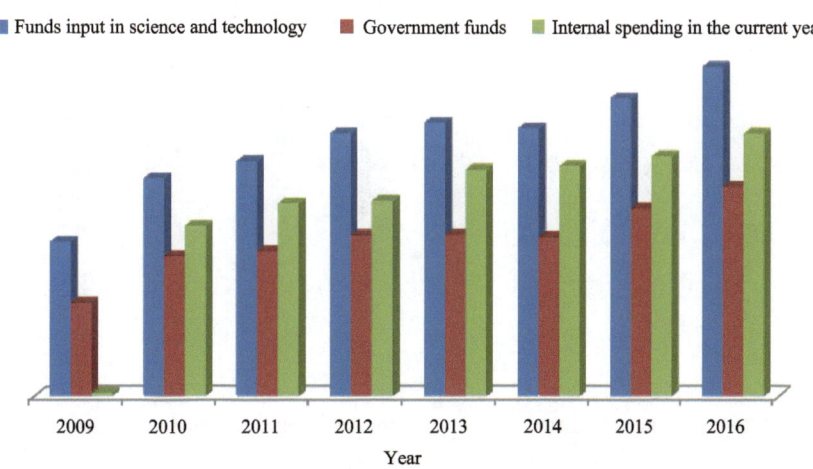

Figure 1-35 Input and Expenditures (Thousand Yuan) of Science and Technology Funds at Marine-related Higher Institutions of China from 2009 to 2016

From 2009 to 2016, the total number of S&T projects at the marine-related higher institutions of China has shown a gradual increase, with an average annual growth rate of 5.80% and the number of staff involved in S&T projects in the current year remained stable (Figure 1-36), with an average annual growth rate of 1.00%. From 2009 to 2016, the funds input and expenditure for S&T projects at the marine-related higher institutions in China in the current year have shown an overall growth trend (Figure 1-37), with the average annual growth rate of funds input in the current year reaching 10.35% and the average annual growth rate of expenditure in the current year reaching 11.06%.

1.5.3 Gradual increase in the marine innovation output of higher institutions

From 2009 to 2016, the number of academic papers published regarding the S&T achievements at China's marine-related higher institutions has gradually increased, with an average annual growth rate of 5.10%, of which the number of academic papers published in foreign academic journals has witnessed an obvious growth, with the average annual growth rate of 13.27% (Figure 1-38). The period of 2009-2010 witnessed the most rapid growth in the number of technology transfer contracts with an annual growth rate of 67.60% (Figure 1-39), followed by a slight decrease from 2010 to 2011 and then followed by a rebounding increase each year with an average annual growth rate of 15.24%.

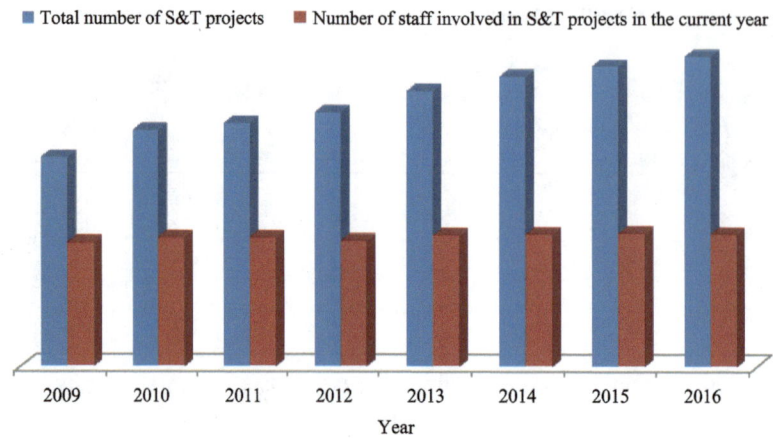

Figure 1-36　Total Number of S&T Projects and the Number of Staff Involved in S&T Projects in the Current Year at Marine-related Higher Institutions of China from 2009 to 2016

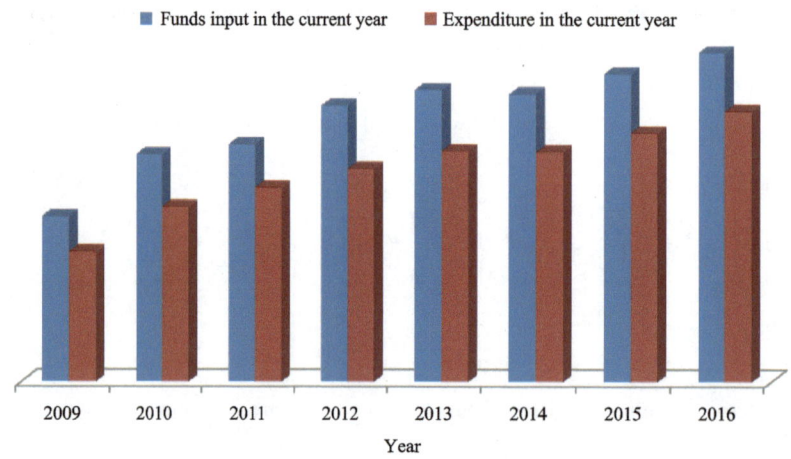

Figure 1-37　Trend of Funds Input (Thousand Yuan) and Expenditure (Thousand Yuan) for S&T Projects at the Marine-related Higher Institutions of China in the Current Year from 2009 to 2016

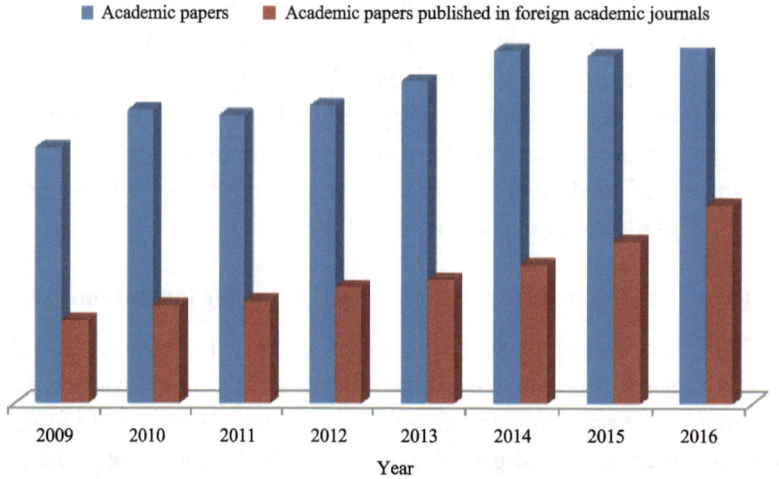

Figure 1-38　Growth Trend of the Number of Academic Papers Published Regarding the S&T Achievements at Marine-related Higher Institutions of China from 2009 to 2016

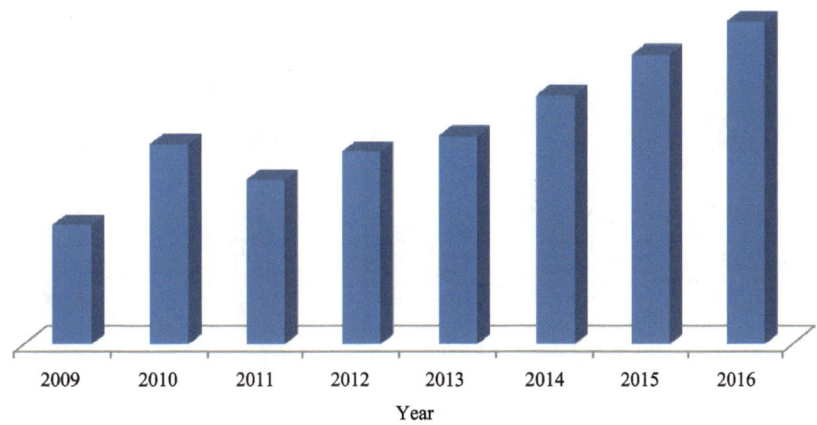

Figure 1-39 Number of Technology Transfer Contracts at Marine-related Higher Institutions of China from 2009 to 2016

1.5.4 Steady development in the marine-related scientific research institutes of higher institutions

From 2012 to 2016, the number of employees at marine-related scientific research institutes of higher institutions in China has witnessed growth each year (Figure 1-40). Of them, the number of doctoral and master graduates has also presented a growing trend; meanwhile, the proportion of doctoral graduates increased from 51.76% to 54.40%, and that of master graduates increased from 27.56% to 27.86% (Figure 1-41).

From 2012 to 2016, the number of S&T personnel at marine-related scientific research institutes of higher institutions in China has gradually increased (Figure 1-42). Of them, the proportion of senior professional title holders have witnessed a minor decline from 60.00% to 59.01%, the proportion of intermediate professional title holders increased from 28.46% to 30.86%, and the proportion of junior professional title holders decreased from 7.76% to 6.30% (Figure 1-43).

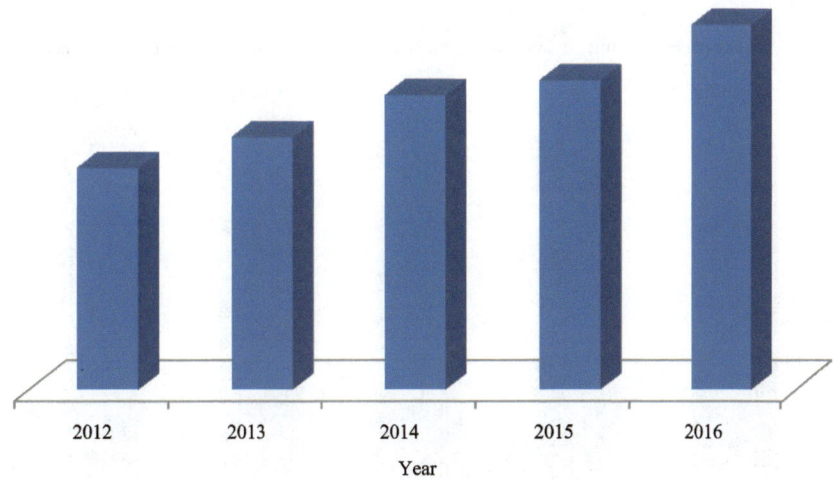

Figure 1-40 Growth Trend of the Employees at Marine-related Scientific Research Institutes of Higher Institutions in China from 2012 to 2016

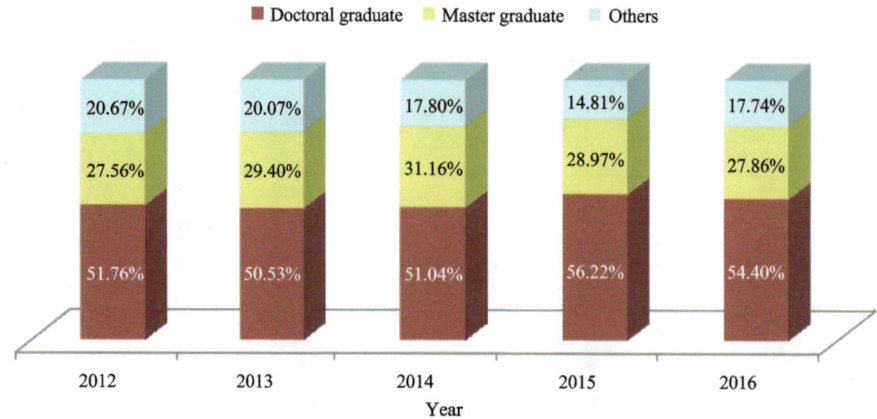

Figure 1-41 Academic Qualification Structure of Employees at Marine-related Scientific Research Institutes of Higher Institutions in China from 2012 to 2016

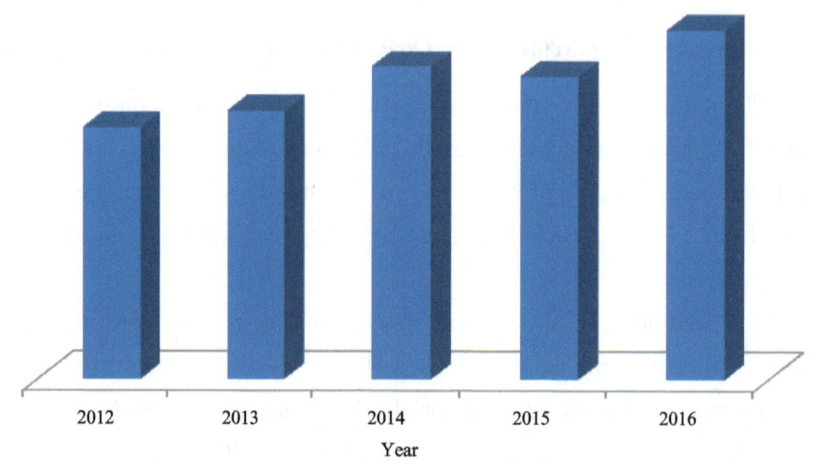

Figure 1-42 Trend of S&T Personnel at Marine-related Scientific Research Institutes of Higher Institutions in China from 2012 to 2016

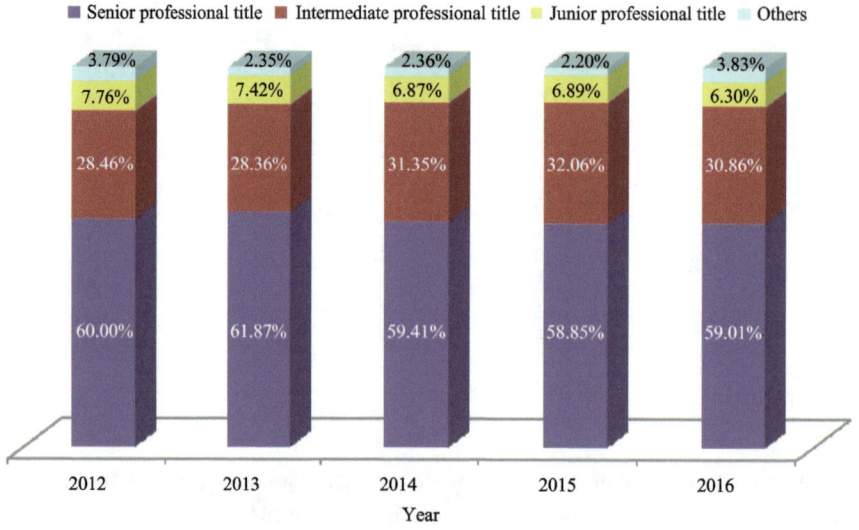

Figure 1-43 Professional Title Structure of S&T Personnel at Marine-related Scientific Research Institutes of Higher Institutions in China from 2012 to 2016

From 2012 to 2016, the expenditures of S&T funds at marine-related scientific research institutes of higher institutions in China have gradually increased (Figure 1-44). The internal spending in 2016 increased by 88.14% compared to 2012, and the R&D expenditure in 2016 increased by 108.02% compared to 2012.

From 2012 to 2016, the total number of project undertaken by marine-related scientific research institutes of higher institutions in China has shown a slow but steady growth trend (Figure 1-45), increasing by 54.00% in 2016 compared to 2012.

From 2012 to 2016, the original value of fixed assets at marine-related scientific research institutes of higher institutions in China has risen each year (Figure 1-46), increasing by 52.25% in 2016 compared to 2012. The original value of instruments and equipment increased by 51.26% in 2016 compared to 2012, and the original value of imported instruments and equipment increased by 70.52% in 2016 compared to 2012.

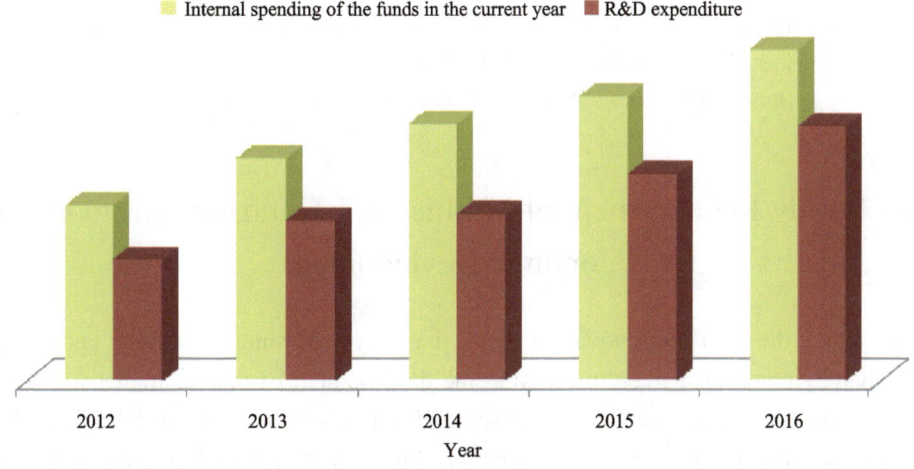

Figure 1-44　The Expenditures of S&T Funds (Thousand Yuan) at Marine-related Scientific Research Institutes of Higher Institutions in China from 2012 to 2016

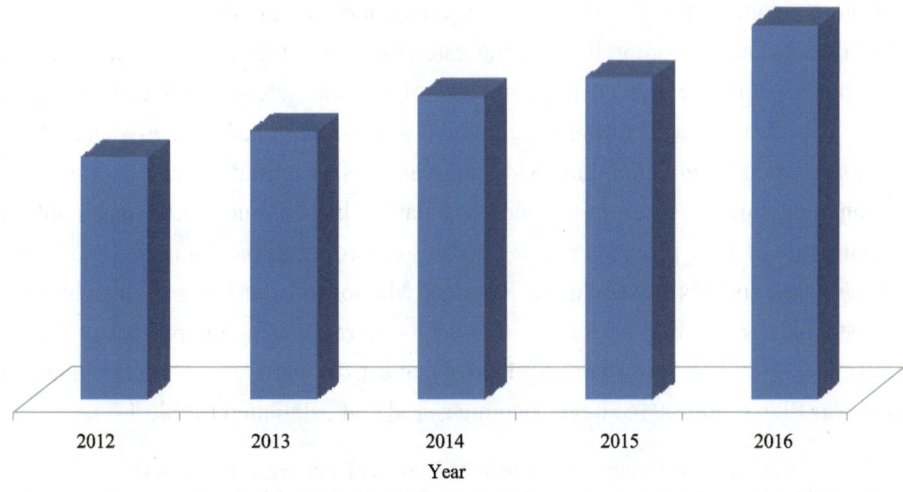

Figure 1-45　Growth Trend of the Number of Projects Undertaken by Marine-related Scientific Research Institutes of Higher Institutions in China from 2012 to 2016

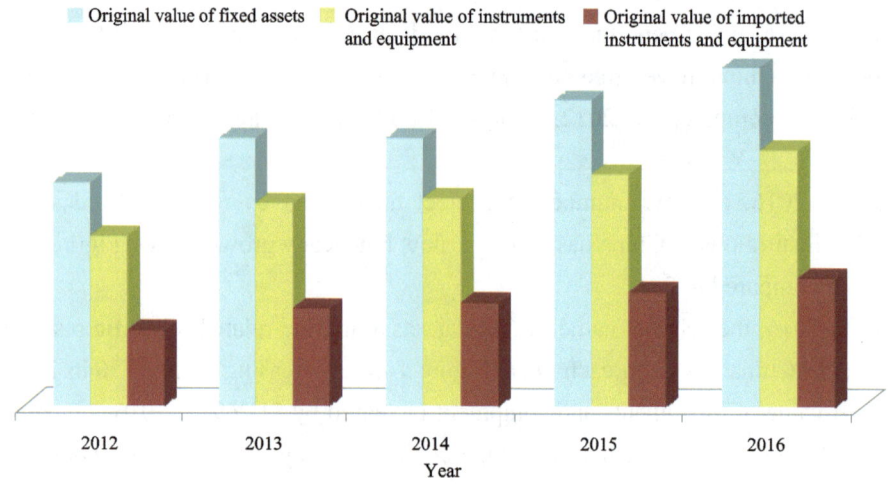

Figure 1-46 Growth Trend of the Original Value of Fixed Assets (Thousand Yuan) Including the Original Value of Instruments and Equipment and Imported Instruments and Equipment (Thousand Yuan) at Marine-related Scientific Research Institutes of Higher Institutions in China from 2012 to 2016

1.6 Steady Enhancement of Marine S&T Contributions to Marine Economic Development

In recent years, the marine innovation work has been advancing solidly and realized staged achievements, thereby comprehensively promoting the development of marine industry. The capability of marine technology to serve marine economic and social development has constantly increased, and the role of technological innovation has become more pronounced in promoting transformation of achievements.

The contribution rate of marine S&T progress witnesses a steady and promising growth. The contribution rate of marine S&T progress refers to the share of the contribution made by marine S&T progress to marine economic growth. It is not only an important indicator for measuring the degree of marine S&T progress, but also a comprehensive indicator for measuring the competitive strength of marine science and technology and the transformation level of marine science and technology into practical productive forces. It is clearly stated in the development goal of *National 13th Five-Year Special Plan for Scientific and Technological Innovation* that "S&T innovation is an important aspect of economic work and plays a more prominent role in promoting economic balance, inclusiveness and sustainable development, and the contribution rate of S&T progress reaches 60%". According to the data of *China Marine Statistical Yearbook* over the years and improved Solow Residual Method on the basis of weighting (Annex 5 for calculation process), the contribution rates of marine S&T progress during the periods of the 11th Five-Year Plan (2006-2010), the 12th Five-Year Plan (2011-2015), and from the 11th Five-Year Plan to the first year of the 13th Five-Year Plan (2006-2016) have been measured and calculated (Table 1-5).

Table 1-5 Contribution Rate of Marine S&T Progress in China (%)

Year	Growth rate of output	Growth rate of capital	Growth rate of labor	Contribution rate of marine S&T progress
2006-2010	12.86	10.10	4.05	54.4
2011-2015	10.97	6.74	2.72	64.2
2006-2016	10.53	6.00	2.56	65.9

As is shown in Table 1-5, the contribution rate of marine S&T progress during the 11th Five-Year Plan period was 54.4%, increasing to 64.2% from 2011 to 2015. The increasing rate was 65.9% during the 2006-2016 period. In other words, from 2006 to 2016, Gross Ocean Product in China grew at an average rate of 12.67%, of which 65.9% came from the contribution made by marine S&T progress, higher than the objective proposed by *Outline of Marine Development by Means of Science and Technology (2016-2020)*. As the beginning year of the 13th Five-Year Plan, 2016 opened up a new condition for marine innovation and development during the time period within the 13th Five-Year Plan.

The transformation capability of marine S&T achievements shows progress. The transformation rate of marine S&T achievements refers to the percentage of marine S&T achievements in the total application of marine S&T achievements. These marine S&T achievements can realize self-transformation or can be transformed into production. They are at the application/production stage, and have been widely applied. Whether marine S&T achievements can be put efficiently into practical productivity will determine the economic development and growth of a country. Accelerating the transformation of marine S&T achievements into practical productivity, and promoting the application of new technologies and new products are a key link in the progress of marine science and technology. They are also the key to the transformation of the marine economy from extensive development to intensive development. *The 13th Five-Year Plan for National Marine Economic Development* put forward that the transformation rate of marine S&T achievements in 2020 will have reached more than 55%. According to the marine S&T statistics made by the Ministry of Science and Technology, and the registered data about marine S&T achievements, the transformation rate of marine S&T achievements from 2000 to 2016 reached 50% (See Appendix 6 for calculation process). The transformation capability of marine S&T achievements still has plenty of room for improvement.

II. Evaluation of National Marine Innovation Index

National marine innovation index is a composite index, which consists of four sub-indexes: marine innovation resources, marine knowledge creation, marine innovation performance and marine innovation environment. Given the comprehensiveness and representativeness of marine innovation activities and the availability of basic data, this report selects 20 indexes (See Appendix 1 for the indicator system) to reflect the quality, efficiency and capability of marine innovation.

The national marine innovation index has been rising notably and the marine innovation capability has been increasing significantly. If the base value of the national marine innovation index in 2004 was set at 100, then the national marine innovation index in 2016 was 240. The average annual growth rate of the national marine innovation index during the time period 2004-2016 and the 12th Five-Year Plan were 7.58% and 6.63% respectively, maintaining a steady developmental momentum.

The sub-index of marine innovation resources has been on the rise overall, with an average annual growth rate of 8.00% during the time period 2004-2016, among which the average annual growth rate of the indicators "R&D funds input strength" and "R&D personnel input strength" were 10.84% and 11.17% respectively, which were the main forces to stimulate the rising of the sub-index of marine innovation resources.

The sub-index of marine knowledge creation has maintained a robust increase with an average annual growth rate reaching 10.84% during the time period 2004-2016, among which the indicators of "S&T works published in the current year" and "number of invention patent grants per 10 thousand R&D personnel" increased rapidly with their average annual growth rate reaching 13.69% and 14.22% respectively, higher than that of other indicators, becoming the leading factors in boosting the increase of marine knowledge creation.

The sub-index of marine innovation performance has remained in a relatively slow growth trend with an average annual growth rate of only 4.94% during the time period 2004-2016 among the four sub-indexes. Among the six indicators of the innovation performance sub-index, the growth of "marine labor productivity" remained relatively steady with an average annual growth rate of 10.95%, thereby playing an active role in boosting the growth of marine innovation performance.

The sub-index of marine innovation environment has stayed on an upward trajectory with an average annual growth rate of 5.39% during the time period 2004-2016. It made an immense leap forward especially during the period from 2005 to 2009 thanks to the rapid increase of the indicators of "per capita GOP of coastal areas" and "proportion of equipment procurement cost in R&D funds".

2.1 Comprehensive Evaluation of Marine Innovation Index

2.1.1 Overall rise in national marine innovation index

The base value of our national marine innovation index of 2004 was set at 100, and then the national marine innovation index of 2016 reached 240 (Figure 2-1). The average annual growth rate was 7.58% during the 2004-2016 period.

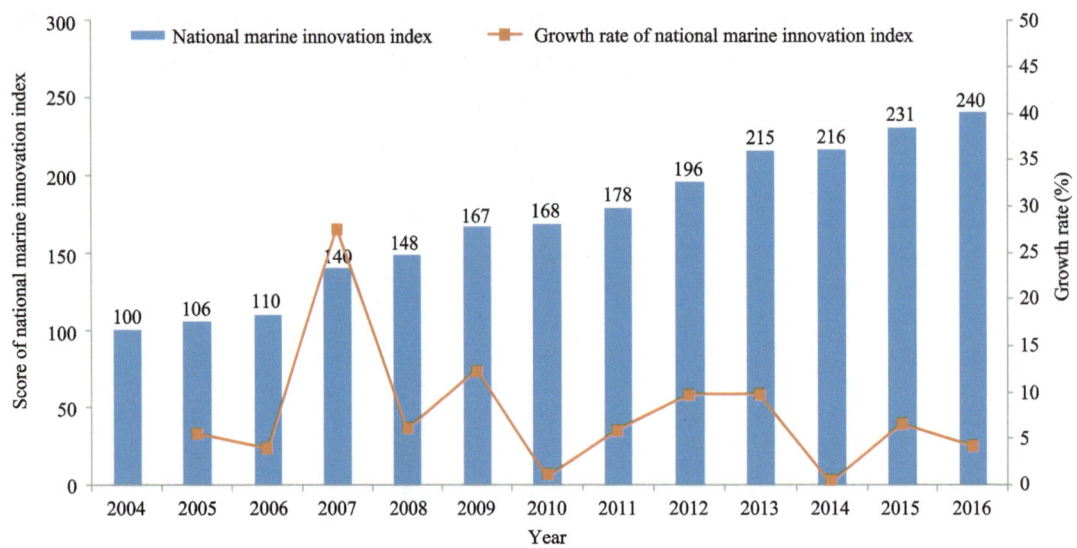

Figure 2-1   Year-on-year Variation of National Marine Innovation Index and Growth rate

The national marine innovation index has generally maintained an upward trend during the 2004-2016 period. The growth speed experienced different degrees of fluctuations. The crest value appeared in 2007 with an annual growth rate reaching 27.52%. The national marine innovation index increased from 110 in 2006 to 140 in 2007. With 2009 as a demarcation point, the national marine innovation index rose rapidly from 2004 to 2009, with an average annual growth rate reaching 10.78%, while the growth rate has slowed down from 2010 to 2016, with an average annual growth rate of 6.11%.

2.1.2 Close relationship between national marine innovation index and the four sub-indexes

The four sub-indexes have different degrees of influence on the national marine innovation index, presenting different upward trends (Table 2-1 and Figure 2-2). The sub-index of marine innovation resources and the variation trend of national marine innovation index were basically the same, both witnessing annual positive growth rate. The scores of the sub-index of marine knowledge creation were substantially higher than that of national marine innovation index, indicating comparatively greater positive contribution of marine knowledge creation to national marine innovation index. The sub-index of marine innovation performance basically presented a steady and slow linear growth trend in spite of small-scale fluctuations in its annual growth rate, forming a certain gap from the growth rate of national marine innovation index. The sub-index of marine innovation environment has remained closest to the scores and trend of national marine innovation index during the time period 2004-2006. Its value was lower than national marine innovation index during the 2007 to 2016 period, but the variation trend was still very close.

Table 2-1　Variation of National Marine Innovation Index and Its Sub-indexes

Year	Comprehensive index	Sub-indexes			
	National marine innovation index A	Marine innovation resources B_1	Marine knowledge creation B_2	Marine innovation performance B_3	Marine innovation environment B_4
2004	100	100	100	100	100
2005	106	102	111	103	106
2006	110	105	109	112	113
2007	140	162	152	115	130
2008	148	172	164	125	132
2009	167	197	197	127	146
2010	168	199	195	136	144
2011	178	208	214	146	146
2012	196	221	251	153	158
2013	215	236	306	157	162
2014	216	239	288	164	174
2015	231	246	327	169	181
2016	240	252	344	178	188

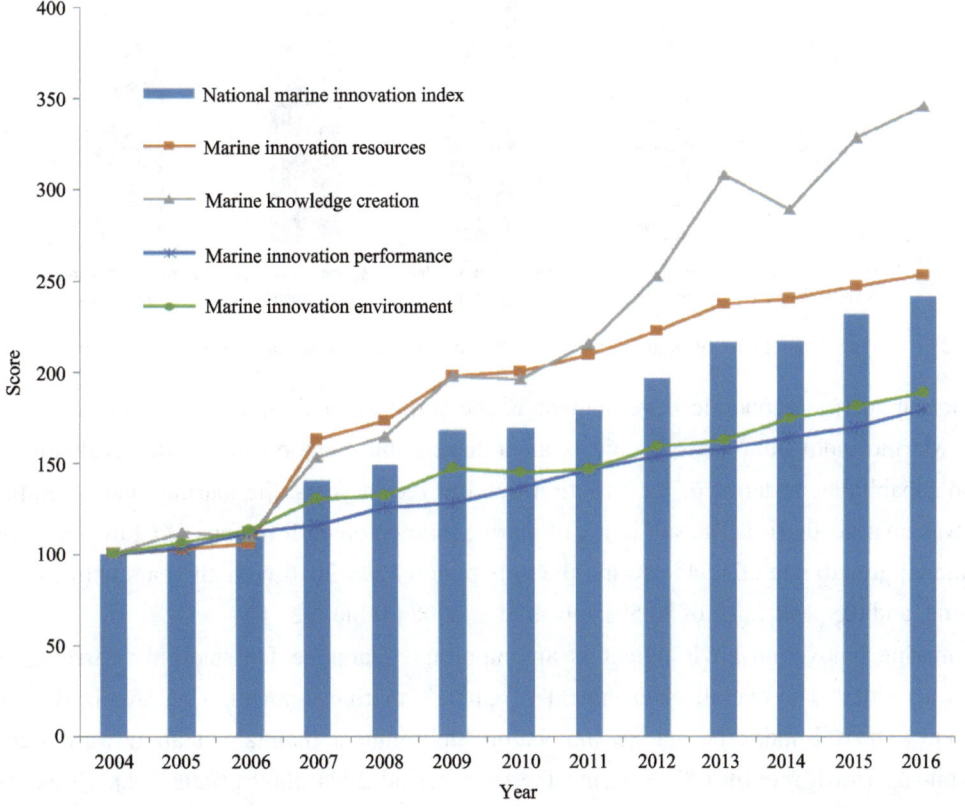

Figure 2-2　Variation Trend of National Marine Innovation Index and Its Sub-indexes from 2004 to 2016

The sub-index of national marine innovation resources has maintained an average annual growth rate of 8.00% during the time period 2004-2016 with all years witnessing positive growth rate, which fully embodied the progressive momentum of continuous investment in the marine innovation resources of China.

From 2004 to 2016, the sub-index of marine knowledge creation witnessed a superior contribution to the significant improvement of the marine innovation capability with an average annual growth rate reaching 10.84% (Figure 2-3), indicating that marine scientific research capabilities have enhanced rapidly and the creation, transformation, and the application of marine knowledge have provided a strong support for marine innovation activities. The improvement of marine knowledge creation capabilities has provided crucial support for strengthening the original national innovation capability in addition to improving the independent innovation level.

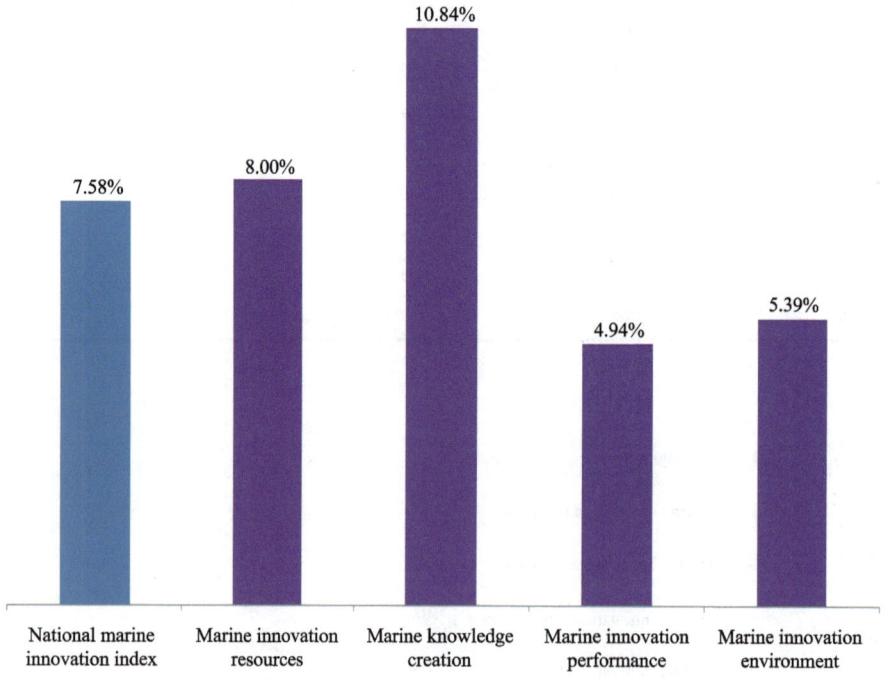

Figure 2-3　Average Annual Growth Rate of National Marine Innovation Index and Its Sub-indexes from 2004 to 2016

Promoting marine economic development is the ultimate goal of carrying out marine innovation activities. Marine innovation performance is an indispensable component for the evaluation of marine innovation capabilities. In terms of the variation trend of recent years, the marine innovation performance of China was on a steady rise. The sub-index of marine innovation performance of China has maintained an average annual growth rate of 4.94% during the time period 2004-2016 with all years witnessing a positive growth trend, and the peak value of 8.45% appeared in 2008 (Table 2-2).

The marine innovation environment is an important guarantee for smoothly carrying out marine innovation activities. The overall environment of China's marine innovation has realized an impressive improvement. The sub-index of marine innovation environment maintained an upward trend with an average annual growth rate of 7.87% during the time period 2004-2009 (Table 2-1). However, the year 2010 saw negative growth for the first time. The average annual growth rate was 5.39% during the time period 2004-2016, ranking third among the four sub-indexes (Figure 2-3).

Table 2-2 Growth Rate (%) of National Marine Innovation Index and Its Sub-indexes

Year	Comprehensive index	Sub-indexes			
	National marine innovation index	Marine innovation resources	Marine knowledge creation	Marine innovation performance	Marine innovation environment
	A	B_1	B_2	B_4	B_5
2004	—	—	—	—	—
2005	5.61	2.18	11.48	3.22	5.58
2006	3.87	3.04	−1.99	8.05	6.79
2007	27.52	53.90	39.36	3.49	15.17
2008	5.93	6.41	7.49	8.45	1.28
2009	12.59	14.30	20.32	1.71	11.08
2010	0.94	0.90	−1.07	6.70	−1.30
2011	5.93	4.65	9.91	7.20	1.10
2012	9.82	6.18	17.36	5.04	8.69
2013	9.91	6.92	21.94	2.60	2.08
2014	0.28	1.05	−6.09	4.20	7.40
2015	6.79	2.94	13.65	3.37	3.92
2016	4.27	2.47	5.14	5.50	3.98
Average annual growth rate	7.58	8.00	10.84	4.94	5.39

2.2 Evaluation of the Sub-index of Marine Innovation Resources

Marine innovation resources can reflect a country's investment in marine innovation activities. The supply capability of innovative human resources and infrastructure input levels that innovation relies on are basic guarantees for the sustainable development of innovation activities. The sub-index of marine innovation resources adopts the following five indicators: R&D funds input strength, R&D personnel input strength, proportion of doctoral graduates in total R&D staff, proportion of S&T staff in total personnel of marine scientific research institutes, and the number of projects undertaken per 10 thousand scientific research personnel. The investment and configuration capabilities of China's marine innovation resources can be evaluated from the perspectives of capital input and human resources input based on the above indicators.

2.2.1 Stable upward trend of the sub-index of marine innovation resources

The sub-index of marine innovation resources in 2016 scored 252 (Table 2-3), increasing slightly compared to 2015. The average annual growth rate during the 2004-2016 period was 8.00%. From the perspective of historical performance, the sub-index of marine innovation resources increased most notably both in 2007 and 2009 when the annual growth rate was 53.90% and 14.30% respectively. After 2010, the sub-index of marine innovation resources has increased year-by-year.

Table 2-3 Scores of the Sub-index of Marine Innovation Resources and Its Indicators

Year	Sub-index: Marine innovation resources B_1	R&D funds input strength C_1	R&D personnel input strength C_2	Proportion of doctoral graduates in total R&D staff C_3	Proportion of S&T staff in total personnel of marine scientific research institutes C_4	Number of projects undertaken per 10 thousand scientific research personnel C_5
2004	100	100	100	100	100	100
2005	102	94	95	112	101	110
2006	105	99	94	120	102	112
2007	162	204	194	174	105	134
2008	172	216	202	198	108	138
2009	197	285	248	207	109	136
2010	199	250	246	238	111	149
2011	208	270	267	251	110	143
2012	221	287	272	286	111	149
2013	236	333	317	281	113	137
2014	239	337	322	275	117	143
2015	246	362	316	289	117	145
2016	252	344	356	286	116	157

2.2.2 Variable characteristics of the indicators

From the variation trend in the scores of the five indicators of marine innovation resources (Figure 2-4), three indicators presented a rapid growth tendency in the later period and two remained essentially flat. Among them, the "R&D funds input strength" revealed the biggest fluctuation, followed by the indicator for "R&D personnel input strength" and "proportion of doctoral graduates in total R&D staff". The above three indicators have maintained an upward trend during the time period 2004-2016, when the average annual growth rate was 10.84%, 11.17% and 9.15% respectively, being the main driving force that gave an impetus to the overall rise of the sub-index for marine innovation resources.

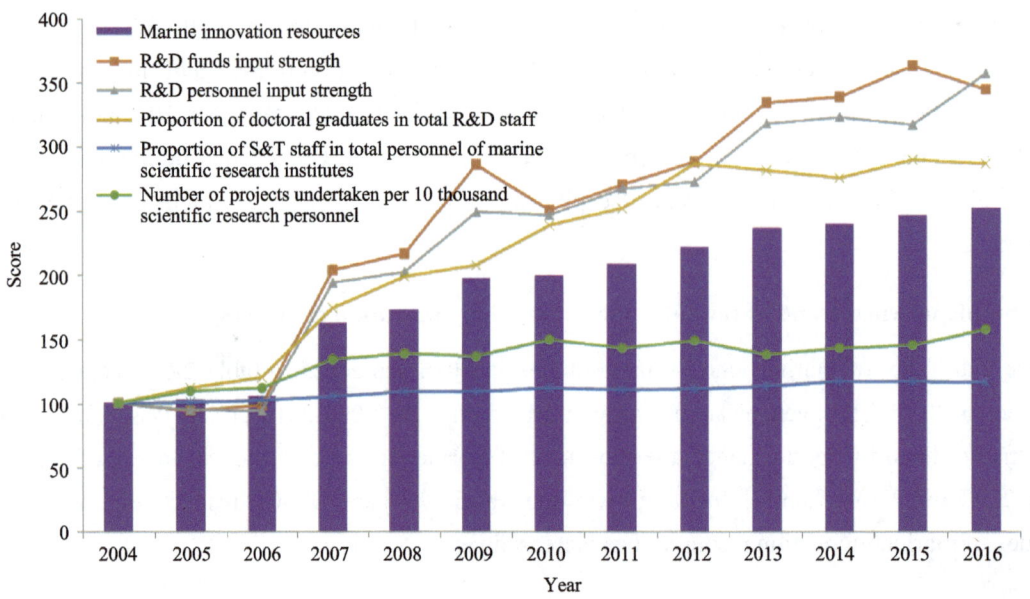

Figure 2-4 Variation Trend of the Sub-index of Marine Innovation Resources and the Scores of Its Indicators

The indicator of the "proportion of S&T staff in total personnel of marine scientific research institutes" can reflect the research strength of a country's marine innovation activities. During the time period 2004-2016, the average annual growth rate was 1.24%, maintaining stability.

The indicator of "number of projects undertaken per 10 thousand scientific research personnel" can reflect the strength of S&T personnel engaged in marine innovation activities. From 2004 to 2008, this indicator maintained a steady upward trend with an average annual growth rate of 8.44%. Negative growth transpired in 2009, and then increased in fluctuations. The scores of this indicator in 2010, 2012 and 2016 were relatively higher. The average annual growth rate was 0.88% during the time period 2010-2016.

2.3 Evaluation of the Sub-index of Marine Knowledge Creation

Marine knowledge creation is the direct output of innovation activities, which can reflect the capabilities of scientific research output and knowledge dissemination within the marine field of a country. The sub-index of marine knowledge creation comprises of the following five indicators: the number of invention patent applications with economic output of 100 million US dollars, the number of invention patent grants per 10 thousand R&D personnel, S&T works published in the current year, the number of S&T papers published per 10 thousand scientific research personnel, and proportion of papers published abroad in total articles. The above indicators provide a full consideration to various achievements such as invention patents, S&T papers and S&T works to demonstrate the capability and level of China's marine knowledge creation.

2.3.1 Increase in fluctuations in the sub-index of marine knowledge creation

The sub-index of marine knowledge creation of China demonstrated an overall upward trend (Table 2-4 and Figure 2-5), increasing from 100 in 2004 to 344 in 2016 with an average annual growth rate standing at

Table 2-4 Scores of the Sub-index of Marine Knowledge Creation and Its Indicators

Year	Sub-index: Marine knowledge creation B_2	Number of invention patent applications with economic output of 100 million US dollars C_6	Number of invention patent grants per 10 thousand R&D personnel C_7	S&T works published in the current year C_8	Number of S&T papers published per 10 thousand scientific research personnel C_9	Proportion of papers published abroad in total articles C_{10}
2004	100	100	100	100	100	100
2005	111	85	124	120	117	110
2006	109	69	142	82	125	129
2007	152	116	118	167	167	193
2008	164	141	145	182	164	187
2009	197	174	172	301	161	177
2010	195	165	178	254	157	220
2011	214	160	238	301	152	219
2012	251	183	319	370	161	224
2013	306	352	298	445	149	290
2014	288	298	327	367	143	304
2015	327	274	475	440	136	311
2016	344	299	493	466	146	315

10.84%. It can be seen from Figure 2-5 that the growth of the marine knowledge creation sub-index can be approximately divided into two stages. The first stage refers to the time before 2013 when marine knowledge creation presented a relatively slow upward trend with an average annual growth rate of 13.25%. The second stage occurs after 2013 when the sub-index of marine knowledge creation fluctuated constantly but tended to present stable trend. The growth rate during the 2013-2016 period was 3.92%. Predominantly benefiting from the indicator "number of invention patent grants per 10 thousand R&D personnel", 2016 witnessed the highest score in marine knowledge creation.

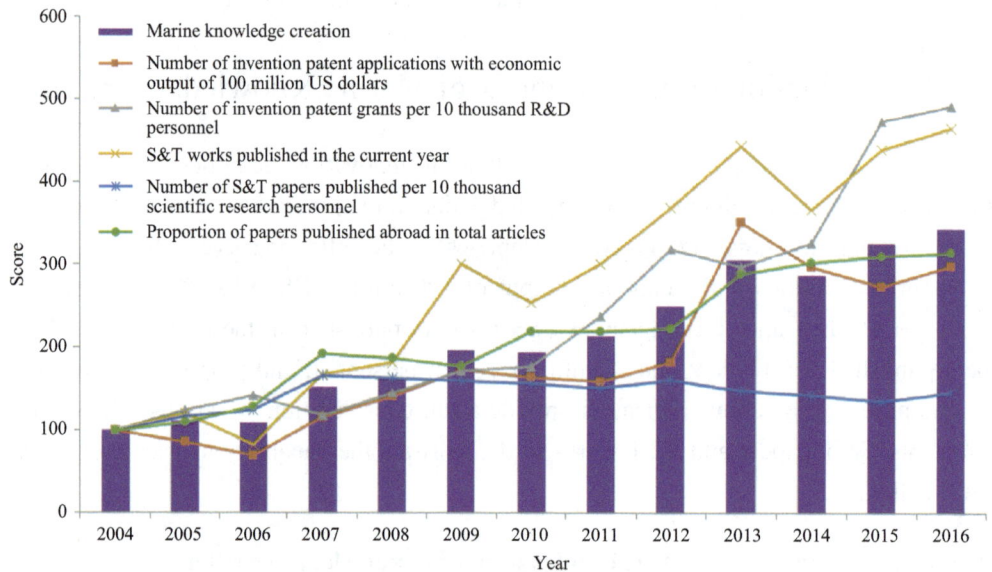

Figure 2-5　Variation Trends of the Scores of the Sub-index of Marine Knowledge Creation and Its Indicators

2.3.2　Different contributions of the indicators

From the variation trend of the five indicators of marine knowledge creation (Figure 2-5), the two indicators of "S&T works published in the current year" and "number of invention patent grants per 10 thousand R&D personnel" fluctuated most dramatically. The indicator of the "S&T works published in the current year" escalated rapidly during the time period 2010-2013, increasing from 254 to 445 with an average annual growth rate of 20.48%. The indicator of the "number of invention patent grants per 10 thousand R&D personnel" also witnessed an impressive rapid growth during the time period 2014-2015, increasing from 327 to 475 with an annual growth rate of 45.23%. In other years, the two indicators showed small fluctuations. Overall, from 2004 to 2016, the two indicators have presented an upward trend in spite of fluctuations, increasing annually by 13.69% and 14.22% respectively.

From 2004 to 2016, the indicator of the "number of invention patent applications with economic output of 100 million US dollars" has presented an upward trend in fluctuations with the average annual growth rate of 9.55%, among which, the negative growth of this indicator appeared during the time period 2004-2006, and then the indicator entered a fast-growth stage after 2007, peaking at 352 in 2013, followed by slight decrease during the time period 2013-2015. The year 2016 scored 299.

"Number of S&T papers published per 10 thousand scientific research personnel", namely the number of S&T papers published by every 10 thousand scientific research personnel, reflects the output efficiency of scientific research. The "proportion of papers published abroad in total articles" refers to the proportion of papers published abroad in total papers published in a country, which can reflect the dissemination degree of the S&T papers to the world. During the time period 2004-2016, the scores of the two indicators rose relatively slowly, with an average annual growth rate of 3.22% and 10.04%.

2.4 Evaluation of the Sub-index of Marine Innovation Performance

The marine innovation performance can reflect the effect and impact of a country's marine innovation activities. The sub-index of marine innovation performance selects the following six indicators: the transformation rate of marine S&T achievements, the contribution rate of marine S&T progress, the marine labor productivity, the proportion of management services of scientific research education in GOP, the marine economic output per unit energy of consumption, and the proportion of GOP in GDP. The effect and impact brought about by the marine innovation activities of China can be reflected by means of the above indicators.

2.4.1 Sequential rise in the sub-index of marine innovation performance

The sub-index of marine innovation performance and the year-on-year scores of the indicators can be seen from Table 2-5. In terms of the scores of the indicators, the sub-index of marine innovation performance of China rose from 100 in 2004 to 178 in 2016, showing a steady and orderly ascending trend with an average annual growth rate of 4.94%, which was the slowest growth rate among the four sub-indexes.

Table 2-5 Scores of the Sub-index of Marine Innovation Performance and Its Indicators

Year	Sub-index	Indicators					
	Marine innovation performance	Transformation rate of marine S&T achievements	Contribution rate of marine S&T progress	Marine labor productivity	Proportion of management services of scientific research education in GOP	Marine economic output per unit energy of consumption	Proportion of GOP in GDP
	B_3	C_{11}	C_{12}	C_{13}	C_{14}	C_{15}	C_{16}
2004	100	100	100	100	100	100	100
2005	103	105	92	113	96	109	104
2006	112	109	108	130	92	122	109
2007	115	112	106	145	91	133	105
2008	125	115	127	165	92	148	103
2009	127	118	119	176	94	153	103
2010	136	120	114	211	85	177	107
2011	146	122	134	238	84	190	105
2012	153	124	143	258	86	201	105
2013	157	126	137	276	87	211	104
2014	164	128	144	305	92	207	104
2015	169	130	145	322	96	219	103
2016	178	129	149	348	106	235	104

2.4.2 Stable trend of indicator variation

"Transformation rate of marine S&T achievements" is an important indicator which can measure the transformation level of marine science and technology into practical productive activities. It has maintained an upward trend from 2004 to 2016 with an average annual growth rate of 2.13%. Overall, the increase of the transformation rate of marine S&T achievements in China was relatively slow before 2010, and has remained stable after 2010 (Figure 2-6).

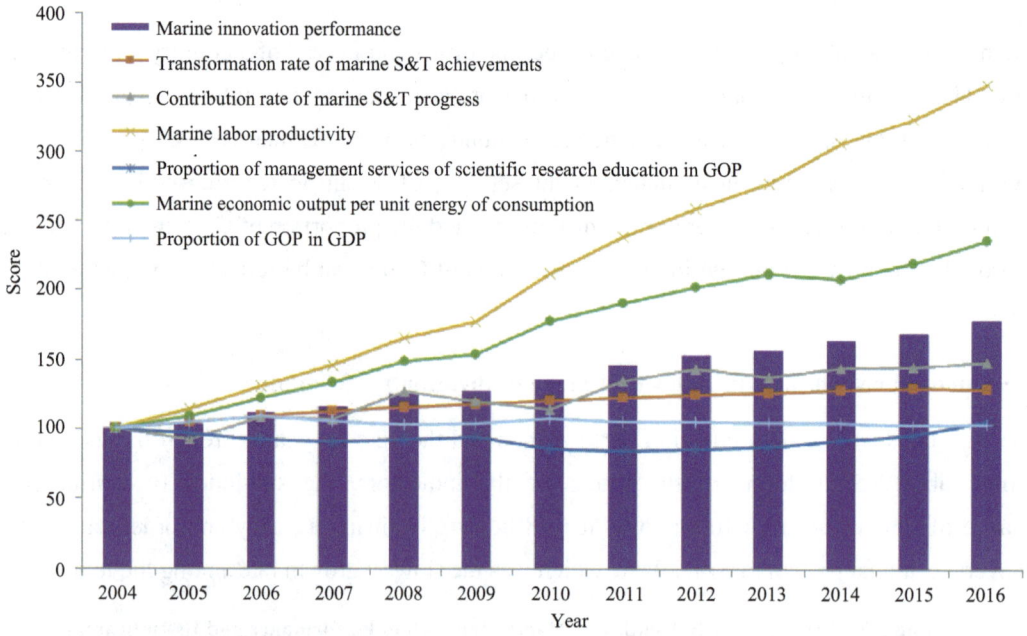

Figure 2-6 Variation Trends of the Scores of the Sub-index of Marine Innovation Performance and Its Indicators

During the time period 2004-2016, the indicator of "contribution rate of marine S&T progress" showed slight fluctuations, yet it still maintained a steady upward trend.

"Marine labor productivity", which refers to the per capita GOP of marine S&T personnel, reflects the role that marine innovation activities play in marine economic output. The indicator of "marine labor productivity" rose rapidly with an average annual growth rate of 10.95% during the time period 2004-2016, which increased most rapidly and steadily among the six indicators of the innovation performance sub-index (Figure 2-6).

The indicator "proportion of management services of scientific research education in GOP" can reflect the contribution of marine scientific research, education and management services to the marine economy. The average annual growth rate during the 2004-2016 period was 0.45%, indicating that the contribution of marine scientific research, education and management services to marine economy showed a relatively rising trend.

The indicator of "marine economic output per unit energy of consumption" uses GOP of energy consumption of 10 thousand tons of standard coal to measure the effect of marine innovation on reducing resource consumption and it also reflects the intensification level of marine economic growth of a country. This indicator increased rapidly during the time period 2004-2016, with an average annual growth rate of

7.40%, showing a steady upward momentum.

The indicator of "proportion of GOP in GDP" reflects the contribution of the marine economy to national economy. It is used to measure the role of marine innovation in boosting marine economy. This indicator showed little change, with a minimal growth rate. The score of 2016 was higher than that of 2004 by only 4, with an average annual growth rate of 0.31% from 2004 to 2016.

2.5 Evaluation of the sub-index of Marine Innovation Environment

Marine innovation environment encompasses both hard and soft environments of the innovation process. It provides an important foundation and crucial guarantee for boosting the innovation capability of China. The sub-index of marine innovation environment reflects the external environment of a country that marine innovation activities rely on, which predominantly refers to institutional innovation and environmental innovation. The sub-index of marine innovation environment selects the following four indicators: per capita GOP of coastal areas, proportion of equipment procurement cost in R&D funds, proportion of government funds in the total science and technology funds of marine scientific research institutes, and full time equivalent workload per capita R&D personnel.

2.5.1 Notable improvement of marine innovation environment

The sub-index of marine innovation environment was on the rise during the time period 2004-2016 (Table 2-6 and Figure 2-7) and the score increased from 100 in 2004 to 188 in 2016 with an average annual growth rate reaching 5.39%. The crest value appeared in 2007, increasing annually by 15.17%, mainly thanks to the rapid increase of the indicator "proportion of equipment procurement cost in R&D funds". The score increased from 105 in 2006 to 151 in 2007.

Table 2-6 Year-on-year Scores of the Sub-index of Marine Innovation Environment and Its Indicators

Year	Sub-index	Indicators			
	Marine innovation environment B_4	Per capita GOP of coastal areas C_{17}	Proportion of equipment procurement cost in R&D funds C_{18}	Proportion of government funds in the total science and technology funds of marine scientific research institutes C_{19}	Full time equivalent workload per capita R&D personnel C_{20}
2004	100	100	100	100	100
2005	106	120	100	99	103
2006	113	143	105	96	107
2007	130	170	151	107	92
2008	132	194	131	107	94
2009	146	209	181	102	93
2010	144	253	128	100	96
2011	146	289	104	99	92
2012	158	326	107	104	97
2013	162	352	94	103	98
2014	174	391	96	106	102
2015	181	414	97	108	103
2016	188	448	99	107	97

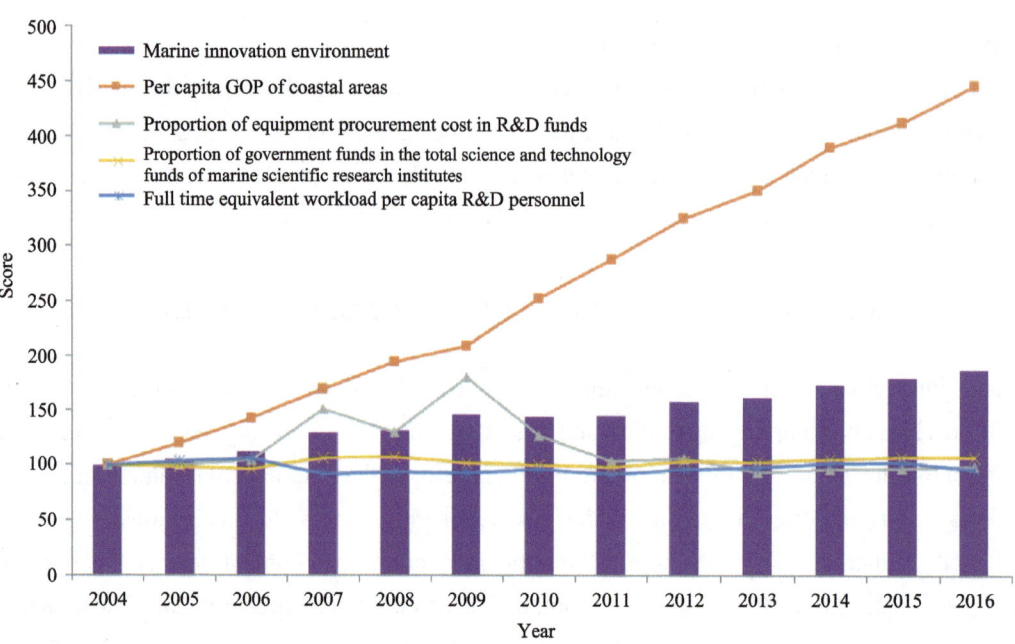

Figure 2-7　Sub-index of Marine Innovation Environment and Variation Trend of Its Indicator Scores

2.5.2　Coexistence of indicators of upward trend with indicators of downward trend

Among the indicators of the marine innovation environment sub-index, the indicator which has always maintained an upward trend was "per capita GOP of coastal areas" with the average annual growth rate at 13.31% and both its score and variation trend remained closest to the marine innovation environment sub-index, increasing the fastest among the four indicators.

While the relatively poorly performed indicators were "proportion of equipment procurement cost in R&D funds" and "full time equivalent workload per capita R&D personnel". In spite of some degree of fluctuations, the score of the "proportion of equipment procurement cost in R&D funds" showed an overall declining trend, cresting in 2009 and then gradually dropping afterwards from 181 in 2009 to 99 in 2016. The score of the "full time equivalent workload per capita R&D personnel" also maintained an overall downward trend, declining from 100 in 2004 to 97 in 2016.

III. Evaluation of Regional Marine Innovation Index

Regional marine innovation is an important component of national marine innovation and its development will influence the pattern of national marine innovation. This chapter analyzes the current situation and characteristics of regional marine innovation and provides a data foundation and decision basis for the optimization of innovation patterns in China.

It clearly stated in the *Vision and Actions on Jointly Building a Silk Road Economic Belt and 21st-Century Maritime Silk Road* that "we should leverage the strengths of the Yangtze River Delta, Pearl River Delta, West Coast of the Taiwan Straits, and Bohai Rim, and other areas with economic zones boasting a high level of openness". By analyzing from the development concepts of the Belt and Road Initiative and looking from the perspective of China's coastal regions, the Chinese people should actively optimize the overall marine economic layout by making use of complementary advantages and practicing joint exploitation. The five economic zones—Bohai Rim, Yangtze River Delta, the West Coast of the Taiwan Straits, Pearl River Delta, and Beibu Gulf Rim economic zones should play a leading role (See Appendix 7 for the definition of marine economic zone) in boosting the formation of three economic circles in the north, east and south of China (See Appendix 7 for the definition of marine economic circle).

In terms of regional marine innovation index of China's coastal provinces[①](autonomous regions, municipalities directly under the central government) (See Appendix 4 for evaluation methods of regional marine innovation index and indicator system), the eleven coastal provinces (autonomous regions, municipalities directly under the central government) of China could be divided into four tiers in 2016.

---

① This evaluation only includes 11 coastal provinces (municipalities directly under the central government, autonomous regions) of the mainland of China which excludes Hong Kong, Macao and Taiwan.

Shanghai and Guangdong belong to the first tier, followed by Shandong and Tianjin as the second tier, Jiangsu, Fujian and Liaoning as the third tier, and Zhejiang, Hebei, Hainan and Guangxi as the fourth tier.

In terms of regional marine innovation index of the five economic zones, the regions with strong regional marine innovation capabilities are the Pearl River Delta Economic Zone, the Yangtze River Delta Economic Zone, and the Bohai Rim Economic Zone in 2016. These areas all have their own regional innovation centers and present a polycentric development pattern.

In terms of regional marine innovation index of the three marine economic circles, the marine economic circles of our country in 2016 presented the following characteristics: the north and east circles were relatively strong while the south circle was relatively weak. The regional marine innovation indexes of the northern and eastern economic circles were higher, reflecting strong original innovation capabilities and fully manifesting the advantages of marine talent agglomeration and the key developing areas of marine economic industries in China.

3.1 Viewing China's Regional Marine Innovation Development from the Perspective of Coastal Provinces (Autonomous Regions, Municipalities directly under the Central Government)

3.1.1 Clear tier characteristics of regional marine innovation

The eleven coastal provinces (autonomous regions, municipalities directly under the central government) of mainland China can be divided into four tiers based on the scores of regional marine innovation index in 2016 (Table 3-1 and Figure 3-1).

Table 3-1 Scores of Regional Innovation Index and Sub-indexes of Coastal Provinces (Autonomous Regions, Municipalities Directly under the Central Government) in 2016

Coastal provinces (autonomous regions, municipalities directly under the central government)	Comprehensive index	Sub-indexes			
	Regional marine innovation index a	Marine innovation resources b_1	Marine knowledge creation b_2	Marine innovation performance b_3	Marine innovation environment b_4
Shanghai	65.06	65.69	47.49	90.91	56.14
Guangdong	61.51	47.88	95.64	57.67	44.85
Shandong	56.50	45.41	56.42	52.90	71.25
Tianjin	54.90	64.40	25.35	70.83	59.03
Jiangsu	49.83	82.47	50.55	38.98	27.32
Fujian	46.69	34.02	31.42	57.95	63.37
Liaoning	44.48	59.68	60.05	31.24	26.95
Zhejiang	36.33	34.28	40.83	36.69	33.53
Hebei	30.13	44.22	29.76	15.04	31.50
Hainan	22.74	8.84	2.57	57.07	22.47
Guangxi	19.71	8.27	13.11	8.36	49.08

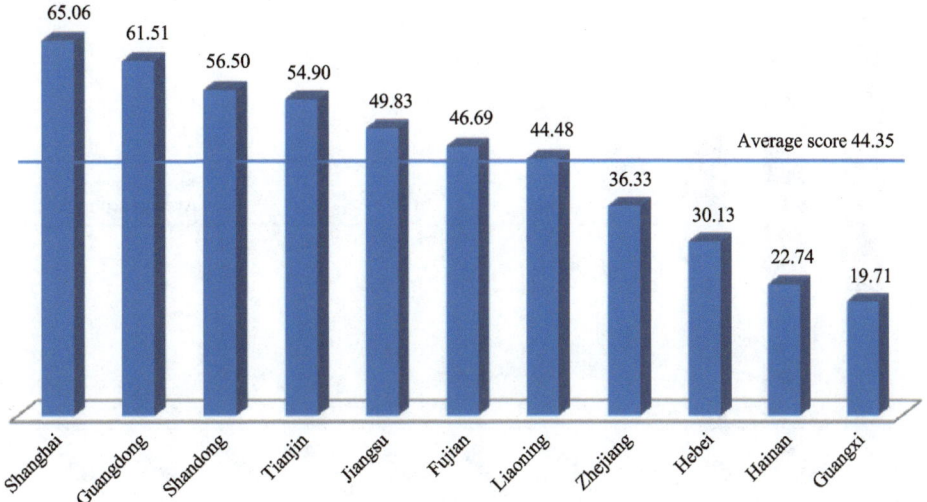

Figure 3-1 Scores of Regional Marine Innovation Index in Coastal Provinces (Autonomous Regions, Municipalities Directly under the Central Government) and Average Score in 2016

In terms of regional marine innovation index, the first tier includes Shanghai and Guangdong. Shanghai scored 65.06, 1.47 times as much as the average score of eleven coastal provinces (autonomous regions, municipalities directly under the central government), ranking first among the total eleven coastal provinces (autonomous regions, municipalities directly under the central government). Shanghai boasts a solid foundation for marine innovation development and shows a strong original innovation capability. The score for Guangdong was 61.51, its ranking rose to the second place from the third place in 2015, thanks to the great improvement of its marine innovation capabilities driven by the rapid development of its marine knowledge creation. The second tier includes only Shandong and Tianjin, with their scores at 56.50 and 54.90 respectively, much higher than 44.35, the average score from the eleven coastal provinces (autonomous regions, municipalities directly under the central government). These two regions have a foundation for marine innovation and have accumulated a large amount of innovative resources over a long period of time. They possess better innovation environments and have made remarkable achievements in innovation performance. Jiangsu, Fujian and Liaoning belong to the third tier, and their scores were 49.83, 46.69 and 44.48 respectively, hovering around the average score. In recent years, the marine economy saw rapid development in these regions, the innovation resources continuously increased, the innovation environment significantly improved, and knowledge creation and innovation performance drastically and rapidly improved. The fourth tier includes Zhejiang, Hebei, Hainan and Guangxi. These provinces' scores were 36.33, 30.13, 22.74 and 19.71 respectively, significantly below the national average. Comparatively speaking, their marine innovation resources were weak, knowledge creation efficiency was not high and innovation environments were in desperate need of improvement.

In terms of the sub-index of marine innovation resources, in 2016, the coastal provinces (autonomous regions, municipalities directly under the central government) with scores above the average score included Jiangsu, Shanghai, Tianjin, Liaoning, Guangdong and Shandong (Figure 3-2). Among them, Jiangsu scored 82.47, far higher than other regions. Shanghai, Tianjin and Liaoning were relatively similar with the scores at 65.69, 64.40 and 59.68 respectively. Shanghai ranked the first among the eleven coastal provinces (autonomous regions, municipalities directly under the central government) both in the input strength of funds and human resources.

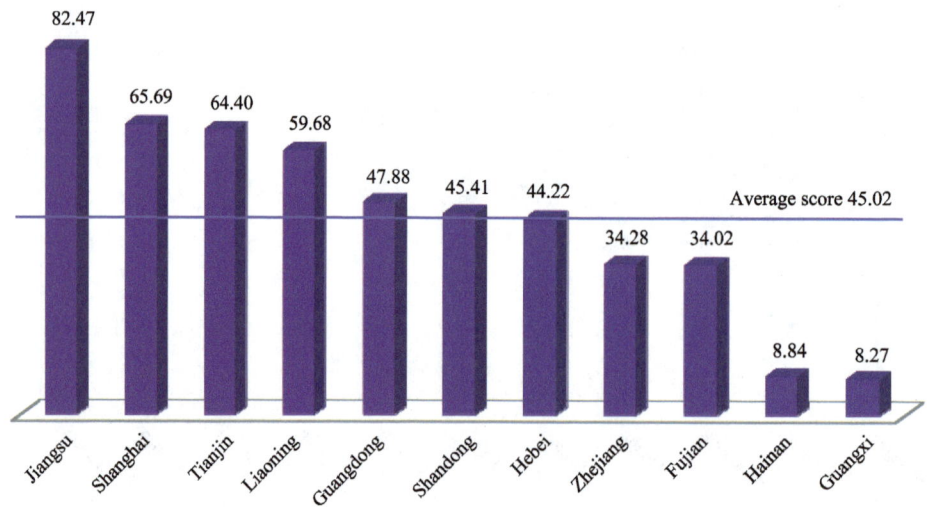

Figure 3-2 Sub-index Scores of Regional Marine Innovation Resources of Coastal Provinces (Autonomous Regions, Municipalities Directly under the Central Government) and Average Score in 2016

In terms of the sub-index of marine knowledge creation, in 2016, the coastal provinces (municipalities directly under the central government) with scores exceeding the average score were Guangdong, Liaoning, Shandong, Jiangsu and Shanghai (Figure 3-3). Among them, the score of Guangdong's regional marine knowledge creation sub-index was an impressive 95.64, far exceeding the average score of 41.20 and increasing by 56.58% compared to last year, which was inseparable from Guangdong's high-output, high-quality marine S&T works and papers. The score of Liaoning's regional marine knowledge creation sub-index was 60.05, mainly benefiting from the marine S&T invention patents. Shandong scored 56.42 primarily due to marine S&T works and papers. Jiangsu scored 50.55, which was mainly attributed to the high-output, the high-quality papers and the patents. Shanghai scored 47.49, mainly benefiting from marine S&T invention patents.

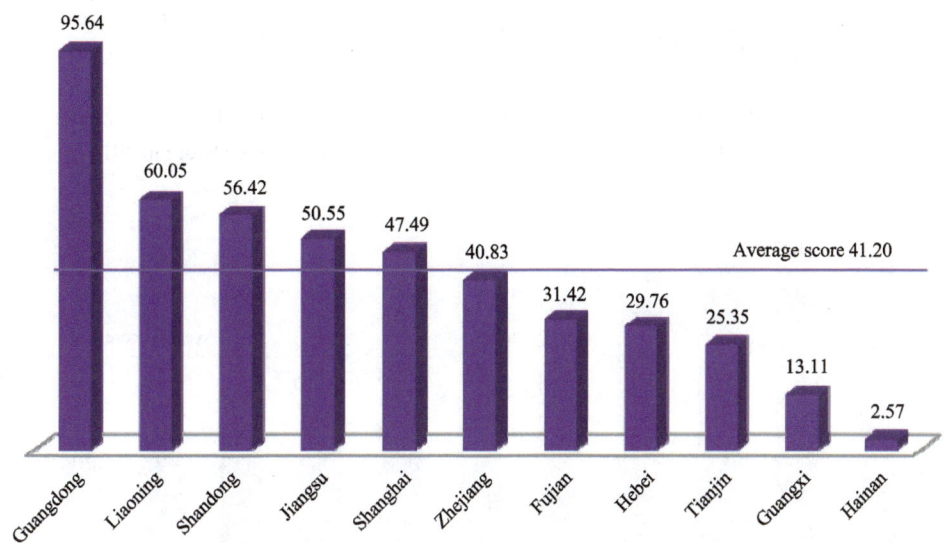

Figure 3-3　Sub-index Scores of Regional Marine Knowledge Creation of Coastal Provinces (Autonomous Regions, Municipalities Directly under the Central Government) and Average Score in 2016

In terms of the sub-index of marine innovation performance, in 2016, the coastal provinces (municipalities directly under the central government) with scores exceeding the average score were Shanghai, Tianjin, Fujian, Guangdong, Hainan and Shandong (Figure 3-4). Among them, the score of Shanghai's regional marine innovation performance index was 90.91, primarily because its labor productivity is abundantly higher than that of other regions and it has a superior marine economic output. Benefiting from its better marine economic output, Tianjin scored 70.83, just slightly below Shanghai. The scores of regional marine innovation performance index of Fujian, Guangdong, Hainan and Shandong were 57.95, 57.67, 57.07 and 52.90 respectively, showing that they perform well in all aspects of marine innovation performance and their overall level is above the national average.

In terms of the sub-index of marine innovation environment, in 2016, the coastal provinces (autonomous regions, municipalities directly under the central government) with scores exceeding the average score were Shandong, Fujian, Tianjin, Shanghai, Guangxi and Guangdong (Figure 3-5). Among them, the score of Shandong's regional marine innovation environment sub-index was 71.25 much higher than that of other areas because of a superior environment for marine innovative talents and government funds.

Figure 3-4 Sub-index Scores of Regional Marine Innovation Performance of the Coastal Provinces (Autonomous Regions, Municipalities Directly under the Central Government) and Average Score in 2016

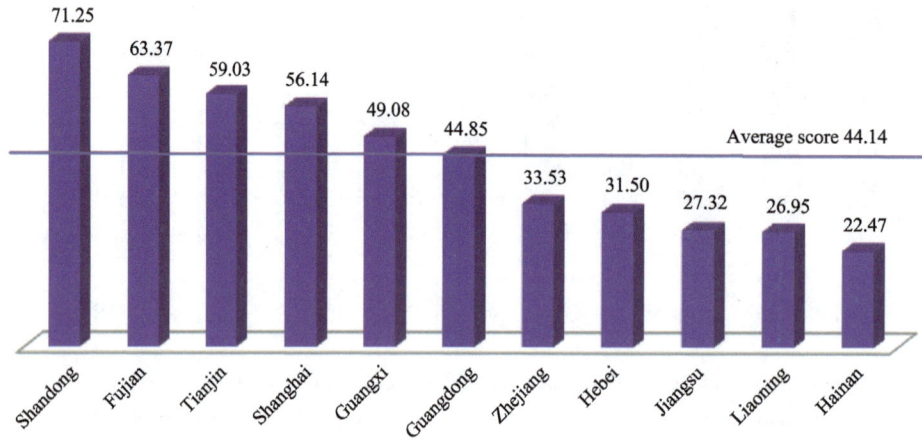

Figure 3-5 Sub-index Scores of Regional Marine Innovation Environment of Coastal Provinces (Autonomous Regions, Municipalities Directly under the Central Government) and Average Score in 2016

Benefiting from state-of-the-art marine equipment and favorable government funding environment, the score of Fujian's regional marine innovation environment sub-index was 63.37. Tianjin scored 59.03 because it boasts better environment for marine innovation funding and a higher per capita Gross Ocean Product. Shanghai scored 56.14 because of higher per capita Gross Ocean Product. The score of Guangxi was 49.08 thanks to a favorable marine innovation funding environment. Guangdong scored 44.85 because it boasts a good government funding environment.

3.1.2 Strong correlation between regional marine innovation capabilities and the level of economic development

Figure 3-6 shows the relationship between "per capita GDP of coastal areas" and the "regional marine innovation index" which is capable of reflecting the level of economic development. The per capita GDP of coastal areas in the first quadrant was relatively high, the regional marine innovation index was higher than the national average, and provinces (cities, districts) in this quadrant belonged to the first and second tier regions; the per capita GDP of coastal areas in the fourth quadrant was relatively high, but the regional marine innovation index was lower than the national average, provinces (cities, districts) in this quadrant belonged to the third tier region except Zhejiang; Hebei, Hainan and Guangxi were in the third quadrant, per capita GDP of which was relatively lower and regional marine innovation index was lower than the national average, belonging to the fourth tier region; there was no region in the second quadrant where the per capita GDP value was lower, but the regional marine innovation index was higher than the national average. All the above data show that there is a strong correlation between regional marine innovation capability and the level of economic development for coastal areas.

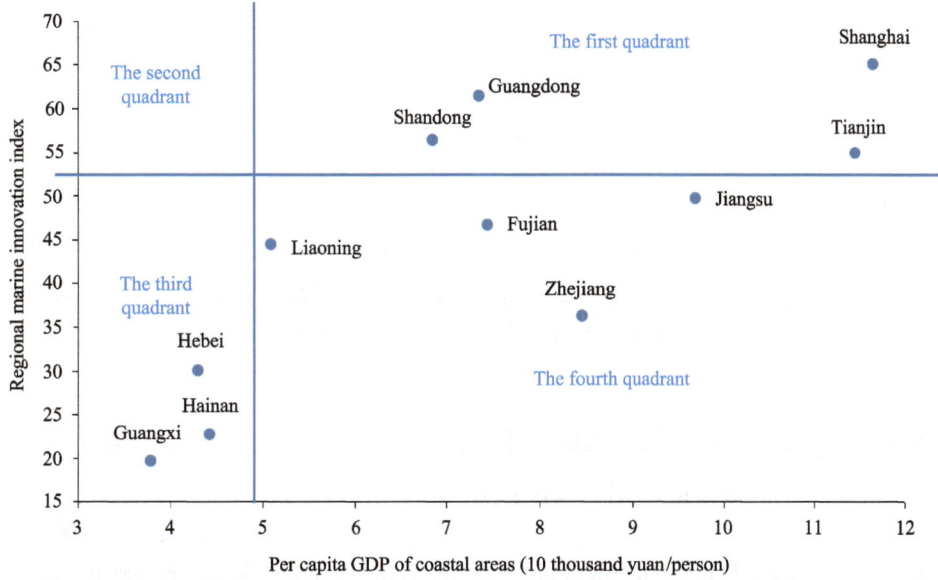

Figure 3-6 Per Capita GDP and Regional Marine Innovation Index of Coastal Provinces (Autonomous Regions, Municipalities Directly under the Central Government) in 2016

Figure 3-7 shows the relationship between "per capita GOP of coastal areas" and "regional marine innovation index". This relationship can reflect the level of marine economic development. The per capita GOP of coastal areas in the first quadrant was relatively high, the regional marine innovation index was higher than the national average, and provinces (cities, districts) in this quadrant belonged to the first and second tier regions. The per capita GOP in the fourth quadrant was relatively high, but the regional marine innovation index was close to or lower than the national average, and provinces (cities, districts) in this quadrant included all regions belonging to the third tier regions and Hainan and Zhejiang which belonged to the fourth tier region. Both the per capita GOP and the regional marine innovation index in the third quadrant were lower than the national average, Hebei and Guangxi in this quadrant belonged to the fourth tier regions. There was no region in the second quadrant where the per capita GOP value was lower than the national average, but the regional marine innovation index was higher than the national average. All the above data show that there is a strong correlation between marine innovation activities and the level of marine economic development of coastal areas.

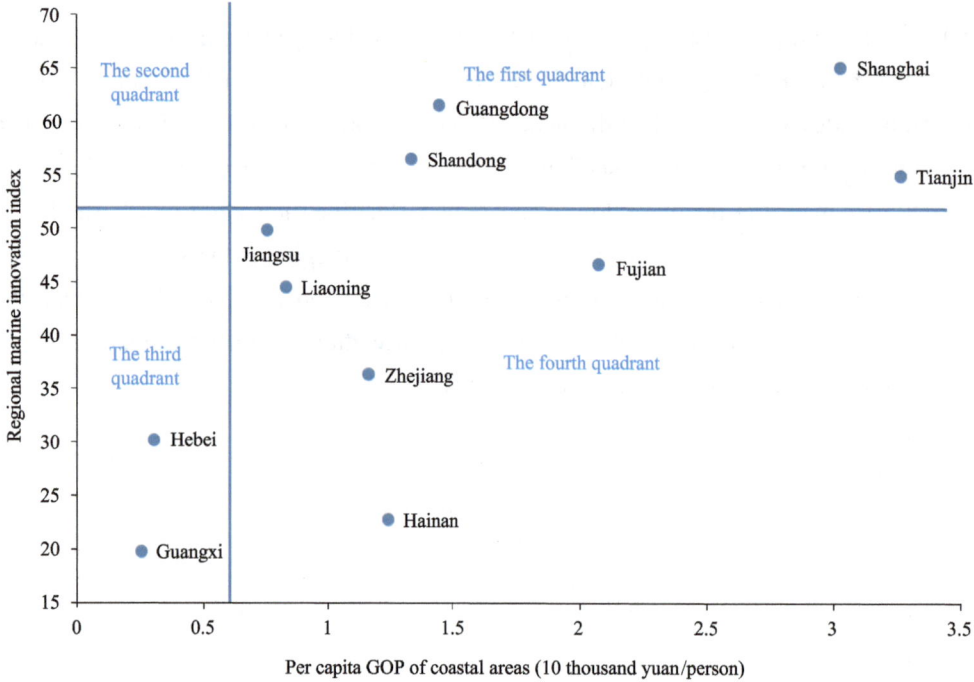

Figure 3-7　Per Capita GOP and Regional Marine Innovation Index of Coastal Provinces (Autonomous Regions, Municipalities Directly under the Central Government) in 2016

3.2　Viewing China's Regional Marine Innovation Development from the Perspective of Five Economic Zones

Specific analysis of the Bohai Rim Economic Zone, Yangtze River Delta Economic Zone, West Coast of the Taiwan Straits Economic Zone, Pearl River Delta Economic Zone, and Beibu Gulf Rim Economic Zone is made as follows.

The Bohai Rim Economic Zone refers to the vast economic areas consisted of the coastal areas surrounding the entire Bohai Sea and parts of the Yellow Sea. It is called the "Golden Coast" in the eastern part of China which possesses a fairly complete industrial base, abundant natural resources, strong S&T strength, and convenient transportation. It is also the strategic area for the development of the central and western regions of China and occupies a pivotal position in the national economic development pattern. In 2016, its regional marine innovation index scored 46.50 (Table 3-2), slightly higher than the average level of the eleven coastal provinces (autonomous regions, municipalities directly under the central government), but the scores of its regional marine innovation performance and marine knowledge creation were below the average, leaving room for improvement in its marine innovation development.

The Yangtze River Delta Economic Zone is located at the intersection of the coastal and riverside regions in the eastern part of China with prominent geographical advantages and strong economic strength. Centered on Shanghai and dominated by technology-based industries, the Yangtze River Delta Economic Zone has strong technical strengths, bright prospects, strong governmental supports, superior environment, good educational development and sufficient talent resources. It is the most dynamic coastal area in China. In 2016, its regional marine innovation index scored 50.41, higher than the average level of the eleven coastal

Table 3-2 Regional Marine Innovation Index and Sub-indexes of China's Five Economic Zones in 2016

Economic zone	Comprehensive index	Sub-indexes			
	Regional marine innovation index a	Marine innovation resources b_1	Marine knowledge creation b_2	Marine innovation performance b_3	Marine innovation environment b_4
Bohai Rim Economic Zone	46.50	53.43	42.90	42.50	47.18
Yangtze River Delta Economic Zone	50.41	60.81	46.29	55.52	39.00
West Coast of the Taiwan Straits Economic Zone	46.69	34.02	31.42	57.95	63.37
Pearl River Delta Economic Zone	61.51	47.88	95.64	57.67	44.85
Beibu Gulf Rim Economic Zone	21.22	8.56	7.84	32.72	35.78
Average value	45.27	40.94	44.82	49.27	46.04

provinces (autonomous regions, municipalities directly under the central government). Vast amounts of marine innovation resources have created favorable conditions for marine S&T and economic development in the Yangtze River Delta Economic Zone, and the marine innovation achievements are remarkable.

The West Coast of the Taiwan Straits Economic Zone has Fujian as its main area, including the neighboring areas. It borders the Pearl River Delta Economic Zone to the south and the Yangtze River Delta Economic Zone to the north and connects with the Taiwan island to the east and Jiangxi's hinterland to the west. It is a regional economic complex with unique advantages and will take the lead in making the national economy to go global. In 2016, its regional marine innovation index scored 46.69, slightly higher than the average of the eleven coastal provinces (autonomous regions, municipalities directly under the central government). The scores of its regional marine innovation environment and performance were both above the national average with a relatively high development potential, but the innovation resources and knowledge creation level were comparatively low and marine innovation development capability has yet to be improved.

The Pearl River Delta Economic Zone mainly refers to Guangdong province in the southern part of the Chinese mainland, bordering Hong Kong and Macao Special Administrative Regions. It boasts exceptional S&T strength, abundant talent resources, and rich marine resources. It is one of the fastest growing regions in the economy. Its regional marine innovation index scored 61.51, which was higher than the average of the eleven coastal provinces (autonomous regions, municipalities directly under the central government), making it rank first among the five economic zones. It has intensive marine innovation resources with a fruitful knowledge for creation and for productive innovation performance.

The Beibu Gulf Rim Economic Zone, the only coastal area of the western development region, is located at the junction of South China's economic circle, southwest economic circle and Association of Southeast Asian Nations (ASEAN) economic circle. It is the region in China connected to the ASEAN countries by both seaway and land borders and has a significant geographical advantage and prominent strategic position. Its regional marine innovation index was only 21.22, far lower than the average level of the eleven coastal provinces (autonomous regions, municipalities directly under the central government) and ranking at the bottom among the five economic zones. Its four sub-indexes of the innovation index were all relatively low, far lagging behind the Yangtze River Delta Economic Zone and the Pearl River Delta Economic Zone.

3.3 Viewing China's Regional Marine Innovation Development from the Perspective of Three Marine Economic Circles

The regional marine innovation index of the eastern marine economic circle was 50.41 in 2016, which ranked at the top among the three marine economic circles (Table 3-3 and Figure 3-8). Among the four sub-indexes, the sub-index scores of marine innovation resources and performance were relatively higher, which stood at 60.81 and 55.52 respectively, making greater positive contributions to marine innovation index for this region and fully demonstrating that this region has an outstanding geographical advantage with strong economic power. Its high-quality marine innovation resources have created favorable conditions for regional marine S&T and economic development. The scores for marine knowledge creation and marine innovation environment were relatively low, which were 46.29 and 39.00, thereby contributing negatively to the marine innovation index of this region (Figure 3-9).

Table 3-3 Regional Marine Innovation Index and Sub-indexes of China's Three Marine Economic Circles in 2016

Economic circle	Comprehensive index	Sub-indexes			
	Regional marine innovation index a	Marine innovation resources b_1	Marine knowledge creation b_2	Marine innovation performance b_3	Marine innovation environment b_4
Northern marine economic circle	46.50	53.43	42.90	42.50	47.18
Eastern marine economic circle	50.41	60.81	46.29	55.52	39.00
Southern marine economic circle	37.66	24.75	35.68	45.26	44.94

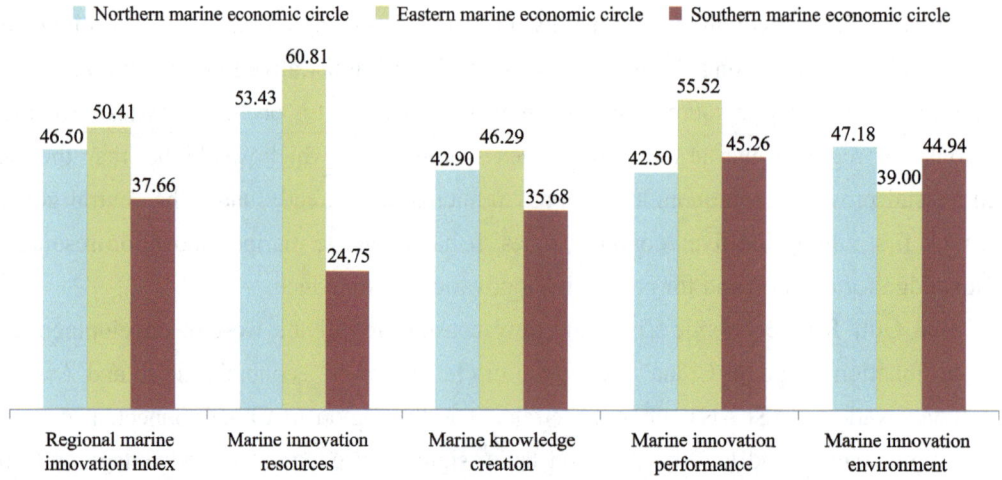

Figure 3-8 Scores of Marine Innovation Index and Sub-indexes of China's Three Marine Economic Circles in 2016

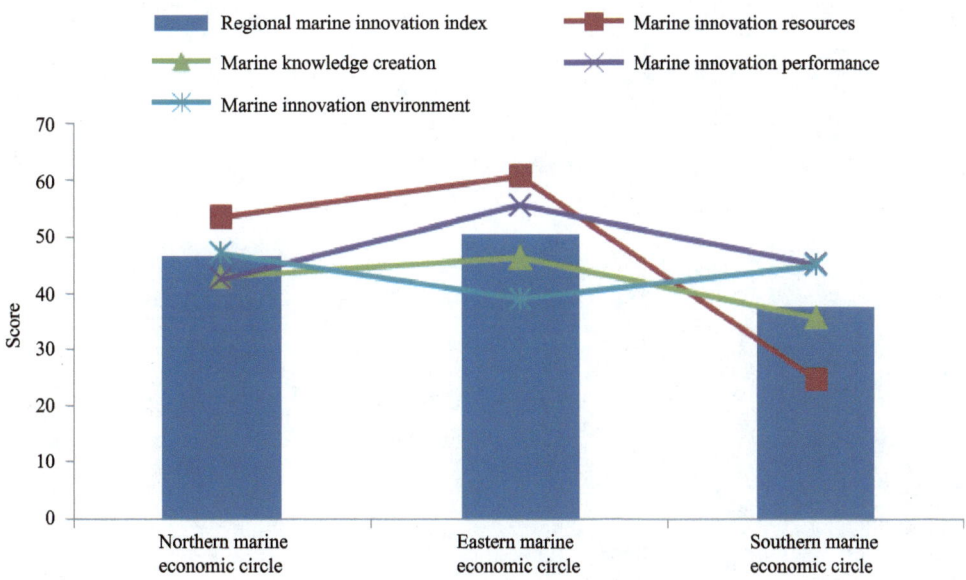

Figure 3-9　Relationship between the Regional Marine Innovation Index and Sub-indexes of China's Three Marine Economic Circles in 2016

The score for the regional marine innovation index of the northern marine economic circle was 46.50, which ranked second among the three marine economic circles. Of the four sub-indexes, marine innovation resources and marine innovation environment contributed positively to marine innovation index, scoring 53.43 and 47.18 respectively. Marine knowledge creation and marine innovation performance scored relatively low at 42.90 and 42.50 respectively. The reason behind the low score for regional marine innovation index in the northern marine economic circle is due to the weak performance on marine innovation, and it is imperative for the development of marine innovation to be further improved.

The regional marine innovation index of the southern marine economic circle scored 37.66, this was the lowest of the three economic circles. The four sub-indexes showed great disparity in scores. Marine innovation environment and marine innovation performance achieved higher scores at 44.94 and 45.26. However, marine knowledge creation and marine innovation resources saw lower scores at 35.68 and 24.75, leading to the lower score for the regional marine innovation index. The lowest score of the southern marine economic circle means that there is considerable room for improvement. In the future process of marine innovation and development, the overall advantages of innovation in the Pearl River Estuary and its two wings should be further utilized to promote the joint development of Fujian, Beibu Gulf and the coastal areas of Hainan Island so as to radiate the mode of marine innovation-driven economic development to the entire southern marine economic circle.

IV. Progress and Prospect of China's Marine Innovation Capability

President Xi Jinping emphasized, "We must develop marine science and technology and promote the transformation of marine science and technology towards a type of innovation leader. We should rely on S&T progress and innovation and strive to break through the technological bottlenecks that hinder the development of the marine economy and marine ecological protection. We must make an overall plan for marine S&T innovation". Innovation is the most important engine leading economic growth. Marine innovation is an important support for guiding the marine industry to make a continuous breakthrough and realize the steady and healthy development of marine economy.

National marine innovation capability and marine economic development complement each other. The improvement of the national marine innovation capabilities is correlated with the development of the marine economy. During the timeframe from 2012 to 2016, the growth rates of national marine innovation index and GOP were close to that of GDP. National marine innovation capability has basically remained consistent with marine economic development and the contribution of marine innovation to economy has enhanced as well.

In the first year of the 13th Five-Year Plan, some indicators of marine S&T development were close to the expected planning targets, and the development trend was positive. In 2016, the proportion of GOP in GDP was 9.51% and the contribution rate of marine S&T progress reached 65.9%, exceeding planned goals. The transformation rate of marine S&T achievements reached 50.0%, which was somewhat lagging behind the planned target of 55%, indicating that there remains room for improvement in the transformation capability of S&T innovation achievements.

4.1 Complementation between National Marine Innovation Capability and Marine Economic Development

National marine innovation capability is complementary to marine economic development. The marine economy provides adequate funds for marine S&T research and development, thus improving the utilization efficiency of marine resources. The progress of marine science and technology and improvement of innovation capability, in turn promotes the growth of the marine economy and the national economy. From 2004 to 2016, the national marine innovation index, GOP, and GDP, all have shown an upward trend (Figure 4-1), with average annual growth rates of 7.58%, 13.98% and 13.63% respectively (Table 4-1). The national marine innovation capability was basically consistent with the trend of marine economic development. However, the growth rate of the national marine innovation index was less than the GOP and GDP, indicating that there still exists plenty of room for improving national marine innovation to contribute to the economic growth.

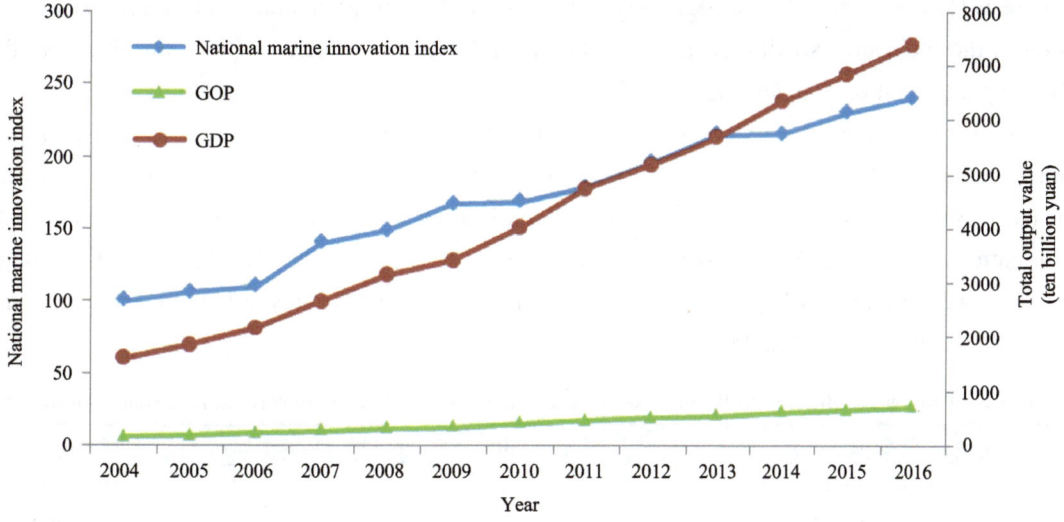

Figure 4-1 National Marine Innovation Index, GOP and GDP from 2004 to 2016

Table 4-1 Growth Rate of National Marine Innovation Index, GOP and GDP (%)

Year	Growth rate of national marine innovation index	Growth rate of GOP	Growth rate of GDP
2004	—	—	—
2005	5.61	20.42	15.67
2006	3.87	22.30	16.97
2007	27.52	18.65	22.88
2008	5.93	16.00	18.15
2009	12.59	8.61	8.55
2010	0.94	22.60	17.78
2011	5.93	14.97	17.83

Continued

Year	Growth rate of national marine innovation index	Growth rate of GOP	Growth rate of GDP
2012	9.82	10.00	9.69
2013	9.91	8.53	9.62
2014	0.28	11.76	11.83
2015	6.79	6.54	7.76
2016	4.27	9.03	8.12
Average annual growth rate	7.58	13.98	13.63

4.2 Progress of the Key Indicators Relevant to the 13th Five-Year Plan National Marine Planning

Both the *Outline of Marine Development by Means of Science and Technology (2016-2020)* and *The 13th Five-Year Plan for National Marine Economic Development* have set clear goals for marine innovation development during the 13th Five-Year Plan period, aiming to guide China's marine innovation development for this period. At the beginning of the 13th Five-Year Plan, doing data analysis of the actual situation of these objectives will be helpful for the management departments of the 13th Five-Year Plan to timely grasp the situation and development trend of national marine innovation capability.

In 2016, the proportion of GOP in GDP reached 9.51% and the contribution rate of marine S&T progress reached 65.9%. The transformation rate of marine S&T achievements reached 50.0% which was somewhat lagging behind the planned target of 55%, indicating that there remains room for improving the transformation capability of S&T innovation achievements. 2016 is the first year of the 13th Five-Year Plan period, and some indicators kept their planning goals and showed an upward trend, indicating a good momentum for development (Table 4-2).

Table 4-2 Progress of the Key Indicators Relevant to the 13th Five-Year Plan National Marine Planning (%)

Key indicators	2015	The goals of the 13th Five-Year Plan	Actual value measured
Proportion of GOP in GDP	9.4	9.5	9.51 (2016)
Contribution rate of marine S&T progress	>60	>60	65.9 (2006-2016)
Transformation rate of marine S&T achievements	>50	>55	50.0 (2000-2016)

Looking into the future, China should further increase investment in marine innovation resources and focus on the efficiency of marine innovation. Play the supporting and leading roles of marine innovation to transform the development mode of the marine economy and promote the transformation and upgrading of marine economy. Rely on marine science and technology to break through the constraints of energy, resources and environment in socioeconomic development, making marine innovation a core force for driving the development, transformation and upgrading of the marine economy so as to provide sufficient knowledge reserve and solid technological foundation for the construction of a strong maritime power.

V. Spatial Distribution Characteristics and Evolutionary Trend of China's Marine Scientific Research Institutions

Marine scientific research institutions are the chief impetus for the development of national marine innovation and an important part for the building of national marine scientific research capability. Under the strategic background of building China into a maritime power, it is of great importance to plan spatial distribution of marine scientific research institutions scientifically and allocate marine scientific innovation resources rationally.

To reveal the distribution of China's marine scientific research institutions and the spatial evolution of the marine scientific research strength, this chapter adopts a standard deviational ellipse method based on single-point data of marine scientific research institutions in the statistics of marine science and technology. It selects practitioners engaged in scientific statistics work, government funds invested in S&T activities, and scientific paper as three indicators to reflect human input, funds input and research output of marine scientific research institutions. This also depicts the spatial-temporal changing process of the geographic location, human input, funds input, research output and other elements of China's marine scientific research institutions from multiple angles. Finally, it reveals the holistic characteristics and dynamic evolution process of China's marine scientific research institutions through spatial visualization, so as to provide decision basis for the formulation of marine S&T innovative development policy and the strategy of distributing marine scientific research institutions as well.

5.1 Research Methods

Standard deviational ellipse (SDE) is a spatial statistical method that can reflect the holistic characteristics of spatial distribution of elements from multiple perspectives. SDE has become a regular statistical tool of ArcGIS spatial statistical module.

SDE method was proposed in 1926 by Lefever, a professor of sociology at the University of Southern California, to measure the direction and distribution of a set of data, and to reveal the spatial distribution characteristics of the elements. The SDE method has been widely used due to its characteristics of being intuitive and effective. The result generated by the SDE method would be an ellipse. According to its generation algorithm, the mean center is used to determine the ellipse center at first, and then the standard deviation of the X and Y coordinates can be calculated with the mean center being the starting point so as to define the axis of the ellipse and determine the direction of the ellipse, due north being 0 degree and rotating clockwise. In addition, the standard deviational ellipse can be calculated according to the position of the elements or the position affected by the attribute value of the associated elements. It should be noted that ArcGIS provides the parameter of the "ellipse size", in which 1, 2 and 3 standard deviational scopes can contain the input elements which can account for about 68%, 95% and 99% of the sum in the ellipses. This chapter selects one standard deviational scope to calculate the geospatial standard deviational ellipses of marine scientific research institutions (hereinafter referred to as the "geo ellipse"), and practitioners, government funds invested in S&T activities, and scientific papers as the attribute value to calculate the weighted standard deviational ellipses of marine S&T institutions (hereinafter referred to as *pe* ellipse, *fi* ellipse and *ot* ellipse).

The SDE method quantitatively describes the spatial distribution characteristics of elements of marine scientific research institutions through the spatial distribution scope of the ellipses and other basic parameters such as center, long axis, short axis and azimuth angle. The connotation of elliptical spatial distribution scope is related to whether it sets the attribute value. For example, the spatial distribution scope of *geo* ellipse directly demonstrates the main geographical area of marine scientific research institutions. *Pe* ellipse is the output ellipse of marine scientific research institutions with practitioners being the weight, and its spatial distribution and the relative position of *geo* ellipse would reflect the spatial distribution of practitioners in marine scientific research institutions. The ellipse center represents the mean center of spatial distribution of elements. The direction of the long axis is the main direction of the spatial distribution of the elements and its length reflects the dispersion degree of the elements in the major trend direction. The short axis shows the scope of spatial distribution of the elements. The higher the ratio of the long axis to the short axis, the more aggregational the data are. Conversely, the dispersion degree is greater. The azimuth angle refers to the angle from the due north in a clockwise direction to the long axis of the ellipse, representing the direction of spatial distribution.

5.2 Spatial Distribution Characteristics and Evolutionary Trends

5.2.1 General overview

The *geo* ellipse, *pe* ellipse, *fi* ellipse and *ot* ellipse of China's marine scientific research institutions from 2001 to 2015 are shown in Figure 5-1. To assist understanding the spatial distribution characteristics and dynamic evolution of China's marine scientific research institutions in terms of geographical location, human input, funds input and output, the standard deviational ellipses of 2001, 2008 and 2015 are selected to produce Figure 5-2 and Figure 5-3.

In terms of geographical location (Figure 5-2), the major trend of the standard deviational ellipses of China's marine scientific research institutions is in the northeastern and southwestern direction, showing a long and narrow distribution. Specifically, in 2001, *geo* elliptical spatial distribution, from the north to the south, covered the following sea areas: Bohai bay, Shandong, Jiangsu, Shanghai, Anhui, Zhejiang, Jiangxi and Fujian Provinces. These areas were the major geographical distribution areas of marine scientific research institutions in 2001. Compared with the *geo* ellipse, the azimuth angle of *pe* and *fi* ellipses was smaller, and the spatial distribution was concentrated to the north because of the apparent advantages in human and funds input of institutions in Beijing. *ot* ellipse had little change in the azimuth angle, and its spatial distribution was centralized in the west, which indicates that the marine scientific research institutions in the north were not dominant in output. In spite of the small number of marine scientific research institutions in the central region (such as Hubei Province), their output was considerable.

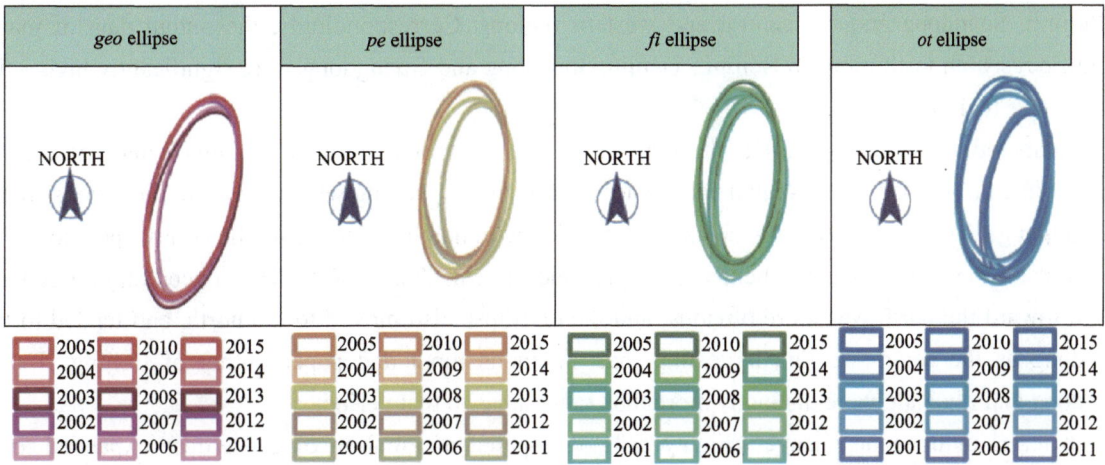

Figure 5-1　Standard Deviational Ellipses of China's Marine Scientific Research Institutions from 2001 to 2015

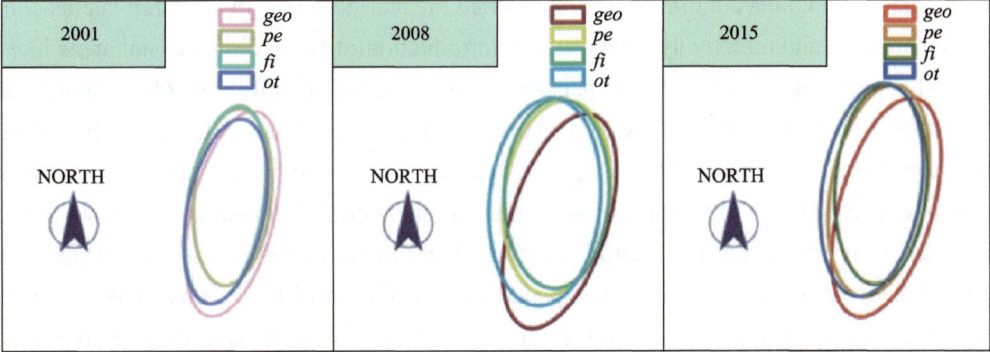

Figure 5-2　Standard Deviational Ellipses of China's Marine Scientific Institutions in 2001, 2008 and 2015 (by Year)

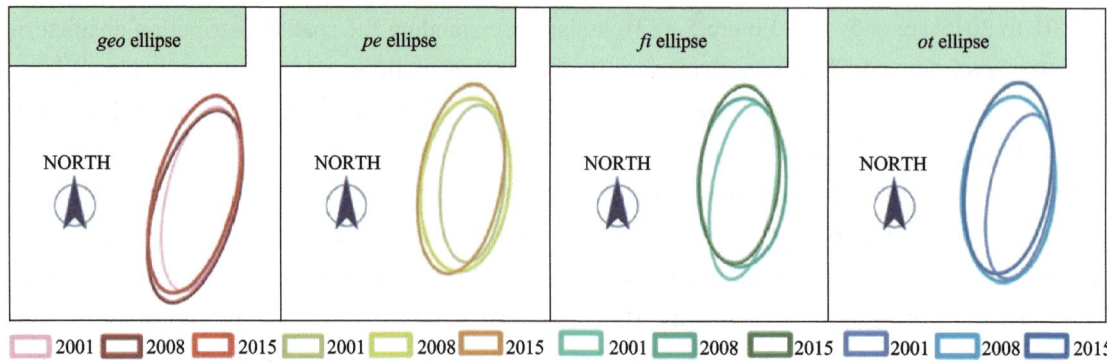

Figure 5-3 Standard Deviational Ellipses of China's Marine Scientific Institutions in 2001, 2008 and 2015 (by Weight)

In 2008, a number of marine scientific research institutions were established in coastal regions such as Guangxi and Hainan, and in central and western regions such as Hubei, Shaanxi and Gansu. Compared with 2001, the *geo* ellipse was extending to the southwest in 2008. In comparison with the *geo* ellipse, the spatial distribution of *pe*, *fi* and *ot* ellipses was centralized to the north, and the azimuth angle became smaller. Among them, *ot* ellipse had the maximum ellipticity, inclining to the west on the whole. The reason was that China significantly increased the human and funds input to marine scientific research institutions in Beijing, Shandong and the central and western regions. Correspondingly, the output data of marine scientific research institutions in Beijing, Tianjin, Shandong and Guangdong were significantly higher than that of other regions.

Compared with 2008, in 2015 the *geo* ellipse moved to the north and China's marine scientific research institutions had a new spatial pattern across the country, among which the northern regions such as Heilongjiang, Liaoning and Beijing increased dramatically in number. Seen from the relative position of the four ellipses, the relation among them was roughly the same as that of 2008. The concentration trend of *fi* ellipse toward the north was more obvious, and the *ot* ellipse also moved to the north, and tended to take Shandong as the central region, indicating that the funds input and paper output of Beijing, Tianjin, Shandong and Zhejiang were distinctly more than other areas in that period.

In terms of dynamic changes (Figure 5-3), the spatial distribution of *geo* ellipse expanded to the southwest first, and then moved to the north. The *pe* ellipse had an expansion as a whole first, moving obviously to the west and south at first, and then moved to the north in the continuous expansion. This expansion indicated that China continuously strengthened the marine scientific research talents construction throughout the country, and effectively promoted the introduction of talents into coastal areas like Guangxi and the central and western regions. The expansion of *fi* ellipse was mainly towards the north, and at the same time it also moved to the west. This indicated that China's marine scientific research investment was still concentrated in Beijing, Tianjin, Shandong, Zhejiang and other provinces (municipalities directly under the central government) that used to be strong in the marine field. Meanwhile, strong support from the government also went to the marine scientific research institutions in the central and western regions. *ot* ellipse showed the most obvious tendency to expand on the whole, just like *fi* ellipse, moving mainly to the north and to the west at the same time. The difference was that the expansion of *ot* ellipse was entirely based on its own, while *fi* ellipse was moving towards the north when expanding, which reflected the nationwide surge of marine scientific research output and indicated the continuous increasing of China's marine scientific research activities and improvement of the marine science and technology innovation.

5.2.2 The variation trend of the mean center

The center of SDE represents the mean center of the spatial distribution of elements. Seen from the trajectory of the center point (Figure 5-4), there was a marked change in the standard deviational ellipse of marine scientific research institutions from 2001 to 2015. In order to summarize the evolutionary rules of marine scientific research institutions, latitude and longitude is not given consideration (Figure 5-5). Figure 5-6 is made in order to clarify the relationship between the geographic location and human input, funds input and research output of marine scientific research institutions.

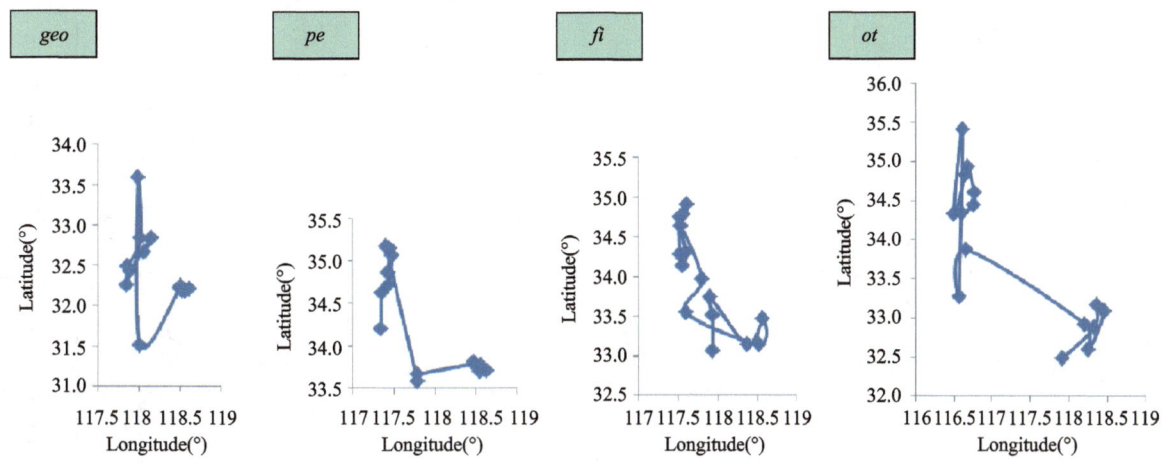

Figure 5-4 Center Point Trajectory of the Standard Deviational Ellipse of Marine Scientific Research Institutions from 2001 to 2015

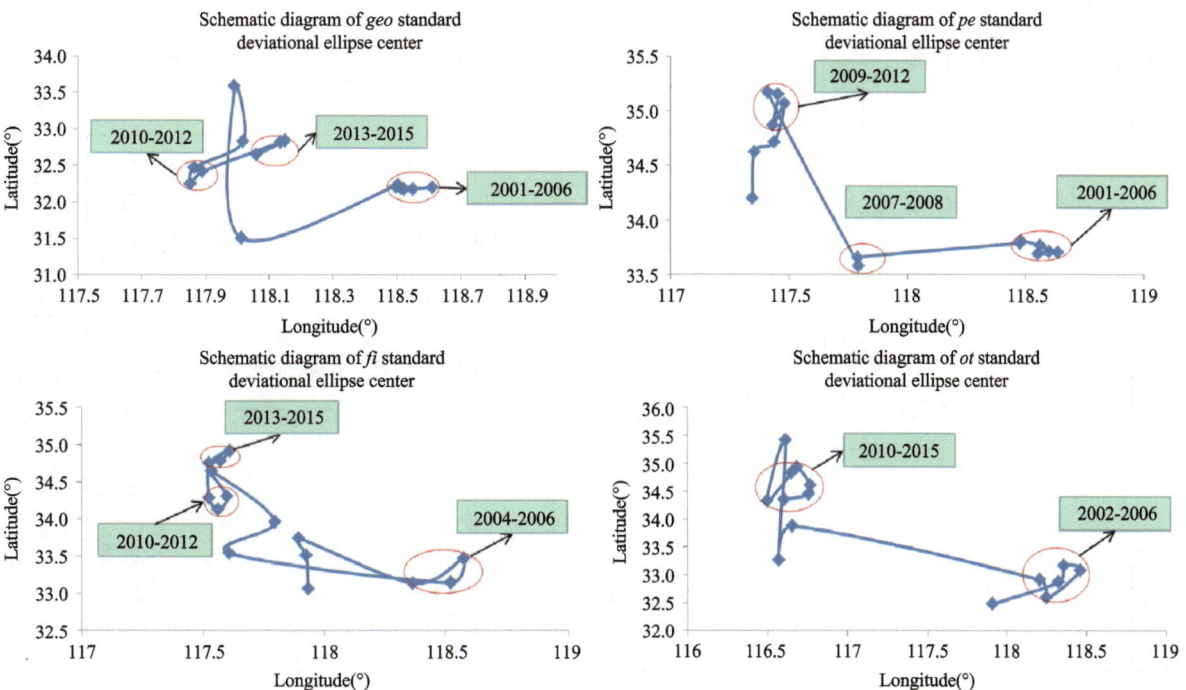

Figure 5-5 Center Point Trajectory of Standard Deviational Ellipse of China's Marine Scientific Research Institutions from 2001 to 2015 (by Weight)

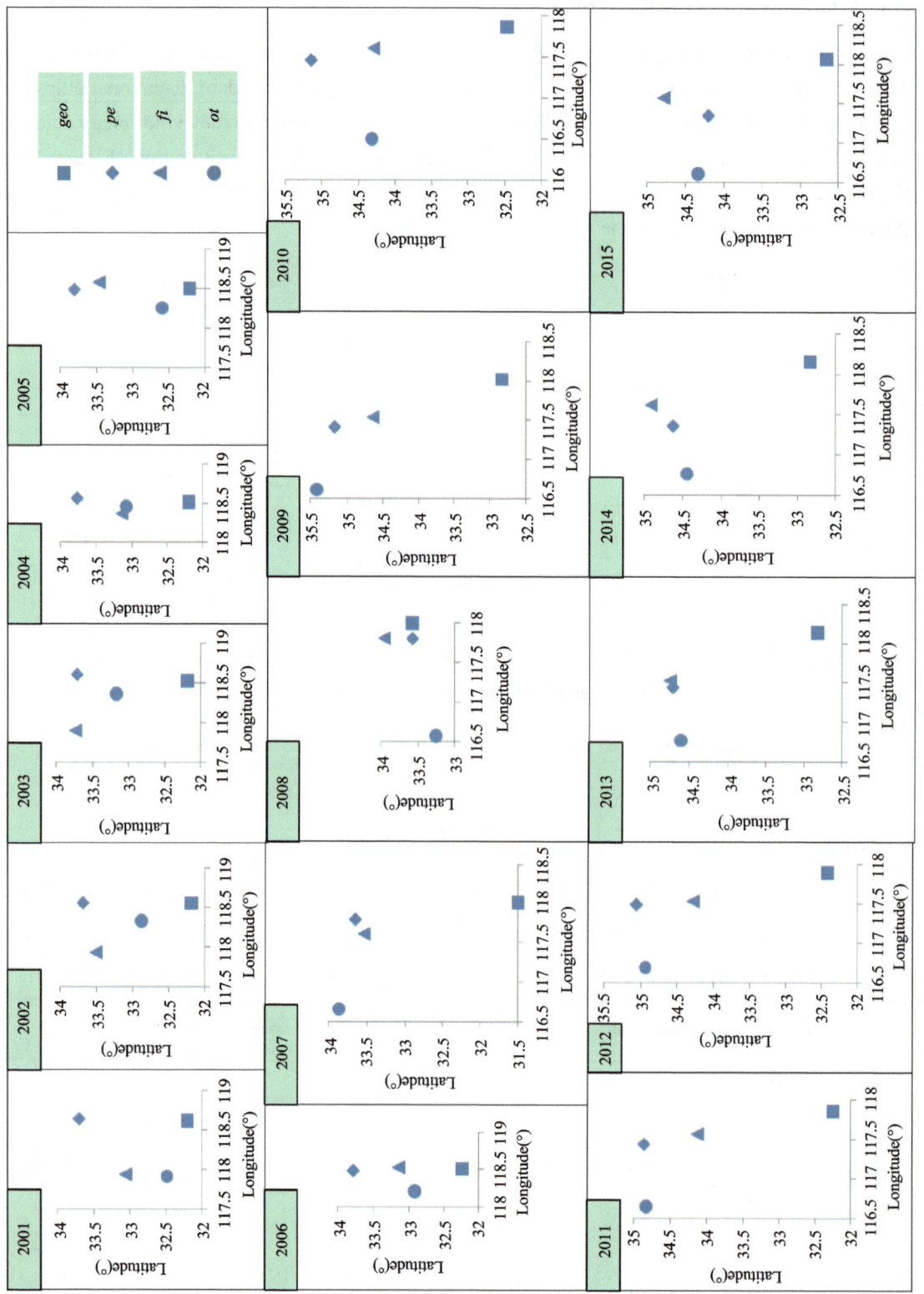

Figur 5-6　Center Point Trajectory of Standard Deviational Ellipse of China's Marine Scientific Research Institutions from 2001 to 2015 (by Year)

As can be seen from Figure 5-4, the center point trajectory of *geo* ellipse can be divided into four phases. The first phase is from 2001 to 2006, during which, most of China's marine scientific research institutions were concentrated in coastal regions with little change. The second phase refers to the years from 2007 to 2009, during which, there was a full-scale outbreak in the construction of marine scientific research institutions. Marine scientific research institutions were established in Heilongjiang, Gansu, Shaanxi, Hubei, Hainan and other places, the number of marine scientific research institutions in coastal regions was also increasing. During this period, the center point of *geo* ellipse significantly changed and moved westward as a whole. The third phase takes place from 2010 to 2012, during which, the center point of *geo* ellipse continued to move westward and was in a stable state. The fourth phase spreads from 2013 to 2015, during which, the number of marine scientific research institutions in Beijing, Tianjin and Shandong increased, and the center point of *geo* ellipse moved northward with a stable tendency.

The center point trajectories of *pe* ellipse, *fi* ellipse and *ot* ellipse need to be explored together with the center point trajectory of *geo* ellipse. For the *pe* ellipse, its center point changed little, and the position which was in the due north of that of *geo* ellipse, remained basically unchanged from 2001 to 2006. This indicated that practitioners of marine scientific research institutions in regions like Beijing, Tianjin, and Shandong had obvious advantages. In 2007, the center point of the *pe* ellipse moved westward and it moved toward the northwest in 2009. From 2013 to 2015, it continuously moved southward, and from 2007 to 2015, it was always in the northwest direction of the center point of *geo* ellipse, indicating that the environment of talents construction for marine scientific research institutions in the central and western regions was favorable. After 2013, the number of practitioners of marine scientific research institutions in the south China has been gradually on the rise. For the *fi* ellipse, its center point was in the northwest direction of the center point of *geo* ellipse throughout. The distance between two center points reduced from 2001 to 2008. In 2009, the distance increased and since then it has remained unchanged. The center point of *ot* ellipse was also in the northwest direction of the center point of *geo* ellipse. From 2001 to 2006, the distance between the two center points fluctuated. In 2007, the distance noticeably increased and since then it has followed a slowly decreasing trend.

The spatial relations of the center points of *pe* ellipse, *fi* ellipse and *ot* ellipse with the center point of *geo* ellipse would provide a reference for determining the balance of the scientific research force in marine scientific research institutions. On the whole, the distance between the center point of *geo* ellipse and that of *pe* ellipse, *fi* ellipse and *ot* ellipse tended to expand, which indicated that the scientific research strength of China's marine scientific research institutions showed a non-equilibrium trend from 2001 to 2015.

5.2.3 The variation trend of the ellipse's long and short axes

As can be seen from the long axis of the standard deviational ellipse of China's marine scientific research institutions (Figure 5-7), the long axis of the *geo* ellipse showed a slow growth trend from 2001 to 2015, indicating that the scope of marine scientific research institutions in the south and north was expanding. The long axes of *pe* ellipse, *fi* ellipse and *ot* ellipse rose in fluctuations, and they were always lower than the long axis of *geo* ellipse, showing apparent clustering of practitioners, government funds invested in S&T activities and scientific papers. From the perspective of the short axis, the short axes of *geo* ellipse, *pe* ellipse, *fi* ellipse and *ot* ellipse has presented upward trend from 2001 to 2015, among which the short axes of *pe* ellipse, *fi* ellipse and *ot* ellipse increased substantially in 2007, demonstrating the fact that the marine scientific research institutions in Heilongjiang, Gansu, Hainan and other regions had significant advantages in the human input, funds input and research output in 2007.

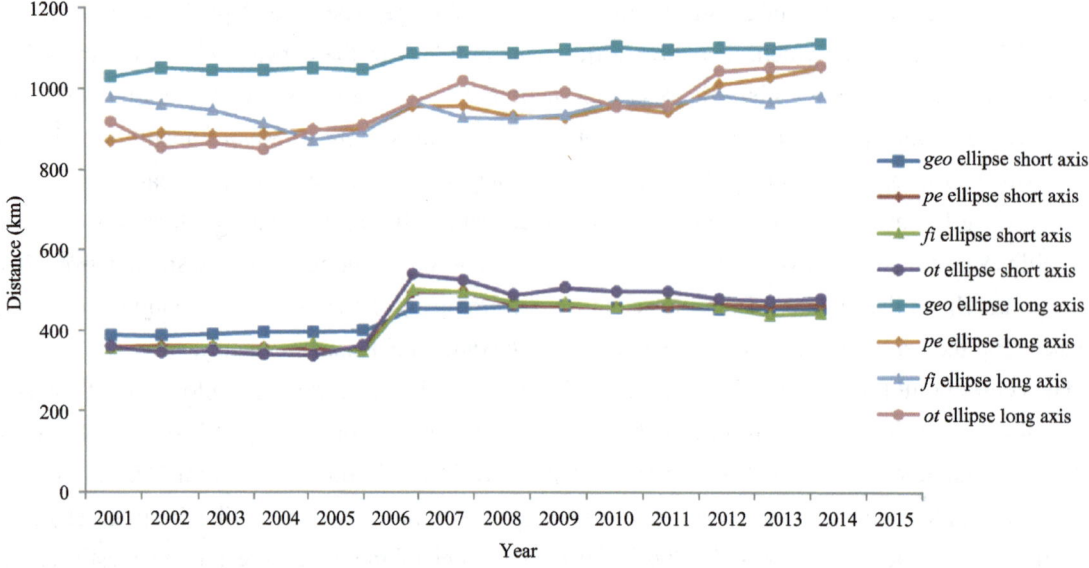

Figure 5-7   Long and Short Axes' Trend of Standard Deviational Ellipses of China's Marine Scientific Research Institutions from 2001 to 2015

In terms of the ratio of the short axis to the long axis (Figure 5-8), the ratio of the short axis to the long axis of the *geo* ellipse grew slowly, while the ratio of the *pe* ellipse, *fi* ellipse and *ot* ellipse rose abruptly in 2007 and declined slowly thereafter. In general, the spatial distribution scopes of China's marine scientific research institutions were expanding in the aspect of both geographical elements and scientific research strength elements, and the dispersion was increasing.

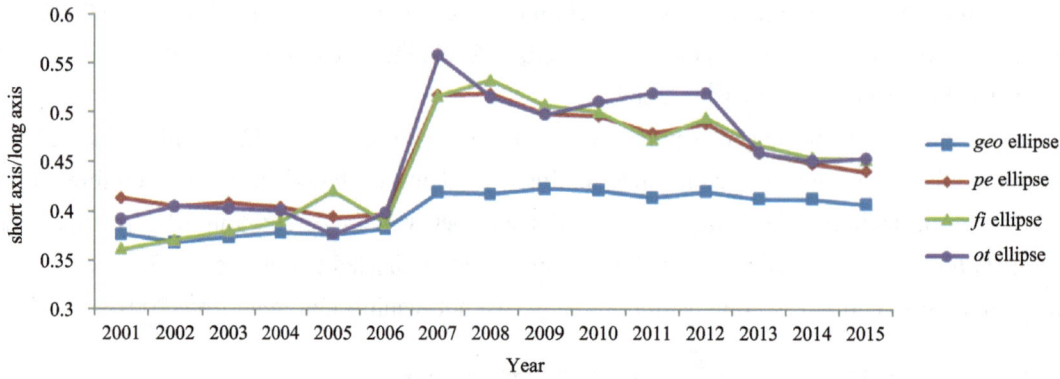

Figure 5-8   Ratio of Short Axis to Long Axis of Standard Deviational Ellipses of China's Marine Scientific Research Institutions from 2001 to 2015

5.2.4   The variation trend of the elliptical azimuth angle

The azimuth angle of the standard deviational ellipse represents the spatial distribution of the elements. As shown in Figure 5-9, the *geo* ellipse azimuth angle changed little during 2001 to 2006. In 2007, the *geo* elliptical azimuth angle increased rapidly, indicating that the number of marine scientific research institutions in the southeast coastal areas increased faster than that in the central and western regions. In

2009, the azimuth angle of *geo* elliptical azimuth became smaller, which reflected the growth trend of marine scientific research institutions in the northern region and the central and western regions inversely surpassed that in the southeast region. From 2010 to 2015, the *geo* elliptical azimuth angle slowly increased, indicating that the marine scientific research institutions in this region maintained their growth advantages in quantity under the aggregation effect of Shandong, Jiangsu, Zhejiang and other established and strong marine provinces.

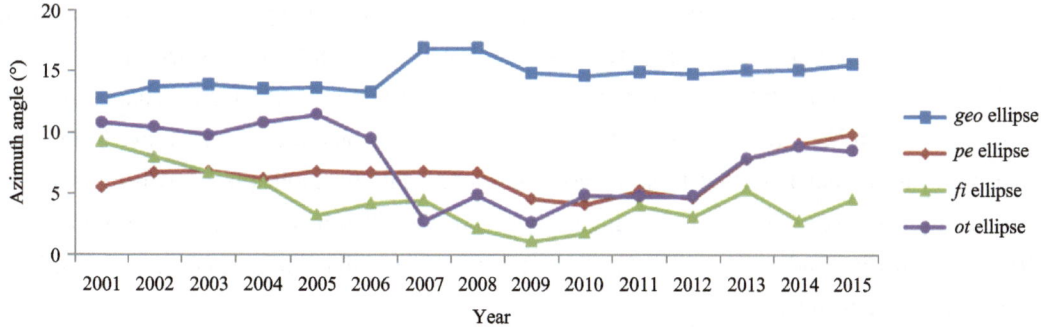

Figure 5-9 North by East Azimuth of Long Axis of Standard Deviational Ellipses of China's Marine Scientific Research Institutions around the Center of Circle from 2001 to 2015

From 2001 to 2008, the azimuth angle of *pe* ellipse remained around 7°, far less than the 12° to 17° of *geo* ellipse, showing the obvious advantages in the number of practitioners in marine scientific research institutions of Beijing, Tianjin and other northern regions. From 2009 to 2010, the azimuth angle of *pe* ellipse became smaller to about 4°, and then continued to become bigger. In 2015, the azimuth angle of *pe* ellipse increased to about 10°, indicating that the advantage in human input of marine scientific research institutions in the northern region was gradually weakened.

The azimuth angle of *fi* ellipse was always smaller than that of *geo* ellipse, showing a downward trend in fluctuations from 2001 to 2015. The investment of the government funds in marine scientific research institutions in north China, such as Beijing and Tianjin, was much higher than other regions. Conversely, the marine scientific research institutions in the central and western regions played a significant role in stimulating funds.

From 2001 to 2015, the azimuth angle of the *ot* ellipse generally showed a U-shape trend. In 2007, the azimuth angle of the *ot* ellipse sharply decreased, indicating that the number of scientific papers published by marine scientific research institutions in the central and western regions, Guangdong, Guangxi and other regions increased rapidly, and reflecting the improvement of the original innovation capacity of marine science and technology in these regions. Since then, the azimuth angle of the *ot* ellipse has increased in fluctuations, and it returned to steady growth after a significant increase in 2013, indicating that the growth trend in the number of scientific papers published in the southeast coastal region has been inversely greater, and the output growth trend of marine scientific research institutions across the country seemed to be convergent.

5.3 Major Research Conclusions

Since the beginning of the 21st century, marine scientific innovation has become the fundamental impetus for marine economic development and it plays a major role in the national and regional competitions. As one of the primary forces of marine S&T innovation, the spatial distribution of marine scientific research institutions' geographic and research elements is of great significance for building China into a maritime power. This chapter reveals the spatial distribution characteristics and dynamic evolutionary process of marine scientific research institutions from 2001 to 2005, and has reached the following conclusions.

A. In terms of geographic location, the number of marine scientific research institutions in China continues to grow and the spatial distribution scope continues to expand. Specifically, most of the marine scientific research institutions are geographically concentrated in coastal areas. Since 2007, a number of marine scientific research institutions have been set up in northern regions like Heilongjiang, in central and western regions like Hubei, Shaanxi and Gansu, and in coastal regions like Guangxi and Hainan etc., but the growth trend is smaller than that in southeast coastal regions. In 2009, the growth trend in northern regions and central and western regions was inversely greater than that in southeastern coastal regions. And after 2010, the growth trend has slowed down, and regions like Shandong, Jiangsu and Zhejiang have restored their growth advantages.

B. In terms of human input, China continues to strengthen the marine scientific research talents construction. Specifically, practitioners of marine scientific research institutions in regions like Beijing, Tianjin and Shandong have obvious advantages. Since 2007, the construction environment of marine scientific research talents in central and western regions has been favorable. After 2013, the human input advantages of the marine scientific research institutions in the north have gradually weakened, and the number of practitioners in the south has been notably increasing.

C. In terms of funds input, China's marine scientific research funds mainly concentrate on old strong marine provinces and municipalities directly under the central government like Beijing, Tianjin, Shandong and Zhejiang. Meanwhile, the marine scientific research institutions in central and western regions play a significant role in stimulating funds.

D. In terms of research outputs, a nationwide growth spurt is now emerging in China's marine scientific research institutions, and the marine research activities have been constantly on the rise. Specifically, before 2007, the marine scientific research institutions in the north did not have an edge in output growth rate, but those in central regions had outstanding performance. In 2007, the number of scientific and technological papers published by marine scientific research institutions in central and western regions, and in regions like Guangdong and Guangxi, increased rapidly. The scientific research outputs in Beijing, Tianjin and Liaoning were also considerable. Since 2009, the growth trend in the number of S&T papers published in southeastern coastal regions has been inversely greater, and the growth trend in output of marine scientific research institutions has been increasing significantly nationwide.

E. In terms of spatial relationship of the four ellipses, the marine scientific research institutions in China show a non-equilibrium trend in geographic location, human input, funds input and research output. From 2001 to 2015, the human input, funds input, and research output all had an obvious discrepancy with the geographical location of marine scientific research institutions in China and the distance from the center points had an expanding tendency, showing an imbalanced developmental trend.

VI. Analysis of Global Marine Innovation Capability

The total number of SCI papers in global marine field maintains a stable growth rate. The number of published SCI papers in 2016 was 1.68 times that of 2001, and the average annual growth rate was 3.54%.

From 2001 to 2016, the top 15 countries with most published SCI papers in global marine field were in order the United States, China, the United Kingdom, Australia, France, Germany, Canada, Japan, Spain, Russia, Italy, Norway, Netherlands, India and Republic of Korea. As for the most frequently cited papers, the United States took the first place, followed by the United Kingdom, Germany, France, Australia and Canada. Although China published the second highest amount of SCI papers, the citation frequency only ranked the seventh.

According to the analysis of the number and annual changes of SCI papers published by the top 20 institutions in global marine field, it was found that eight of the major publishing institutions were from the United States and three were from China. These three Chinese institutions are the Chinese Academy of Sciences, Ocean University of China and original State Oceanic Administration.

The SCI research papers in the marine field not only cover many disciplines but also are interdisciplinary, with the disciplines mainly included but are not limited to oceanographic biology, oceanographic engineering, marine geochemistry, fishery, environment, geology, mining, and other related fields.

EI papers in the marine field showed a rapid developmental trend. The EI papers in the marine field published by China and the United States accounted for 40% of the entire world, and their annual growth rate was much higher than that of other countries. Since 2011, the output of EI papers published in China surpassed the United States, ranking the first in the world.

As for the number of marine patent applications, China ranked first, far ahead of any other country and region. It showed a steady increase in the world share, rising from 5.4% in 2001 to 84.4% in 2016. Among the top 15 institutions with the most international marine patent applications, five were from China, including China National Offshore Oil Cooperation (CNOOC), Ocean University of China, Zhejiang Ocean University, Zhejiang University and Dalian Ocean University.

6.1 Analysis of Total Amount and Situation of Global Marine Innovation Achievements

6.1.1 Stable growth trend of SCI papers

From 2001 to 2016, the total number of SCI papers on global marine innovation was persistent and increased. The number of papers published in 2016 was 1.68 times that of 2001, with an average annual growth rate of 3.54%. As shown in Figure 6-1, from 2001 to 2016, the number of published papers showed a significant climb, which could be divided into three stages, with the years 2006 and 2012 as the turning points.

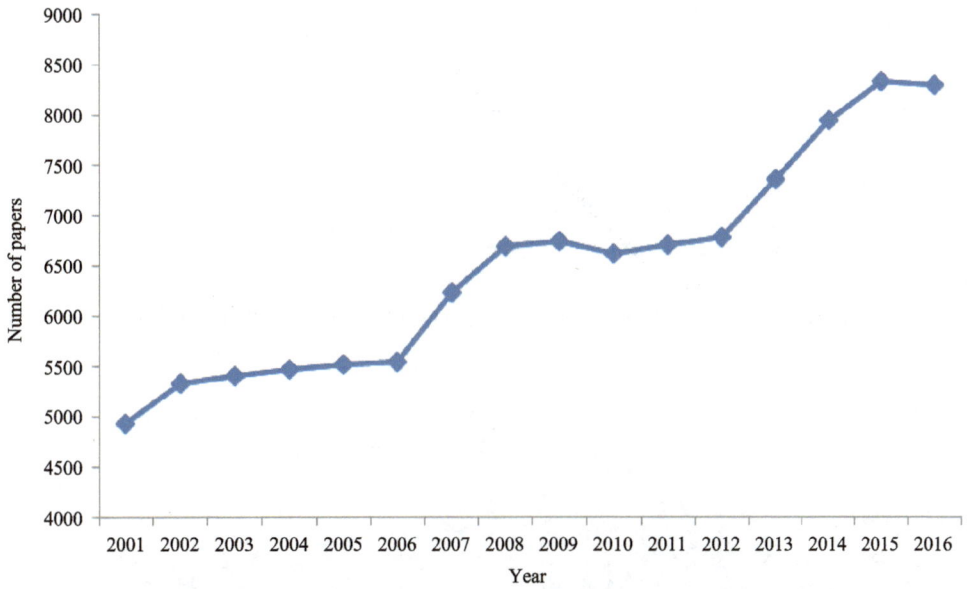

Figure 6-1 Annual Variation in the Number of Published SCI Papers in Global Marine Field from 2001 to 2016

In terms of the number of SCI papers published by the top 20 institutions in the marine field (Figure 6-2), the University of California ranked first, followed by American National Oceanic and Atmospheric Administration (NOAA), Russian Academy of Sciences, Chinese Academy of Sciences, Woods Hole Oceanographic Institution, University of Washington, Ocean University of China, French National Academy of Sciences, University of Hawaii, Ohio State University and other institutions. Among the top 20 institutions, eight were from the United States; three were from China, including the Chinese Academy of Sciences, Ocean University of China, and original State Oceanic Administration. Three were from France with the remaining universities belonging to Canada, Germany, Spain, Japan, Australia and Russia respectively.

Figure 6-3 shows the annual variation in the number of SCI papers in global marine field published by the top 20 institutions from 2001 to 2016. The number of papers published by Chinese institutions in the recent three years took the dominant place. In 2016, the number of papers published by the University of Hawaii, the Spanish National Research Council and the Alfred Wegener Institute for Polar and Marine Research (AWI) and Helmholtz Polar Ocean Research Center were relatively lower.

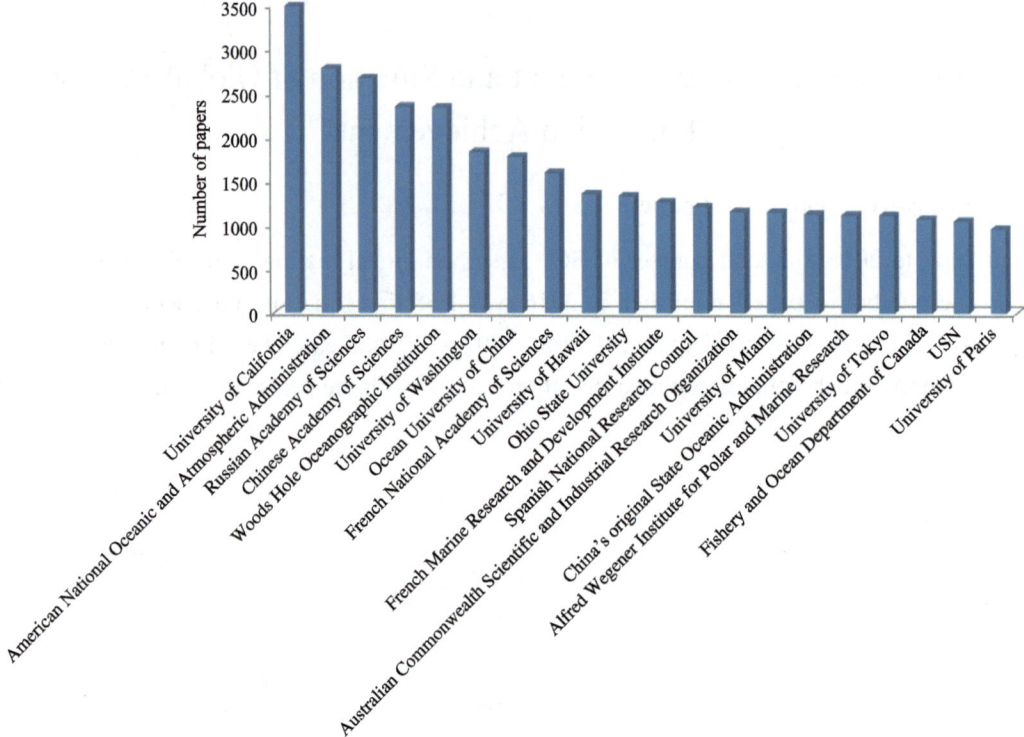

Figure 6-2 Number of SCI Papers in Global Marine Field Published by the Top 20 Institutions from 2001 to 2016

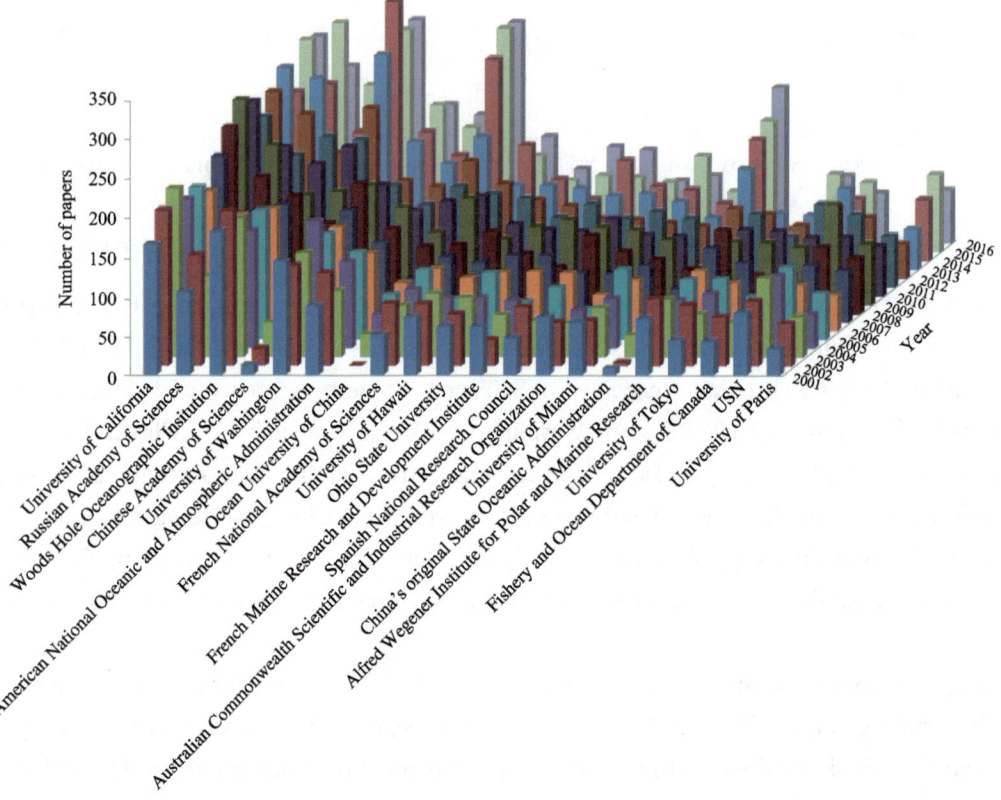

Figure 6-3 Annual Variation in the Number of SCI Papers in Global Marine Field Published by the Top 20 Institutions from 2001 to 2016

Each record in the Web of Science (WOS) database contains a discipline category to which the source publications belong, encompassing 252 discipline categories. Based on the retrieval formula, SCI papers in the marine filed retrieved in the Web of Science database involve 23 disciplines, indicating that marine scientific research comprises many disciplines and those disciplines intersect with each other frequently. The disciplines with the most research results involved are related to oceanographic biology, ocean engineering, marine geochemistry, fishery, environment, geology, mining, and other fields (Table 6-1).

Table 6-1 Global Disciplinary Distribution of SCI Papers in Marine Field from 2001 to 2016

Sequence number	WOS disciplines	Number of papers
1	Oceanography	86 931
2	Ocean Engineering	28 562
3	Marine & Freshwater Biology	28 493
4	Civil Engineering	13 992
5	Meteorology & Atmospheric Science	9 409
6	Ecology	8 916
7	Geosciences, Multidisciplinary	8 284
8	Limnology	7 174
9	Fisheries	6 958
10	Science of Water Resources	4 337
11	Mechanical Engineering	2 485
12	Chemistry, Multidisciplinary	1 535
13	Geochemistry & Geophysics	1 344
14	Paleontology	1 280
15	Electrical & Electronic Engineering	1 153
16	Engineering, Multidisciplinary	858
17	Environmental Science	541
18	Mechanics	541
19	Geological Engineering	429
20	Mining & Mineral Processing	429
21	Remote Sensing	422
22	Zoology	199
23	Energy & Fuel	92

6.1.2 Rapid increase in the number of EI papers

The number of EI papers in the global marine field from 2001 to 2016, as shown in Figure 6-4, was incomplete in recent years due to the time lag in the data collection of papers (the time lag is particularly higher in the collection of conference proceedings and academic papers). From 2001 to 2012, EI papers in the marine research field showed a rapid growth, and the number of papers in 2012 was more than five times that of 2001.

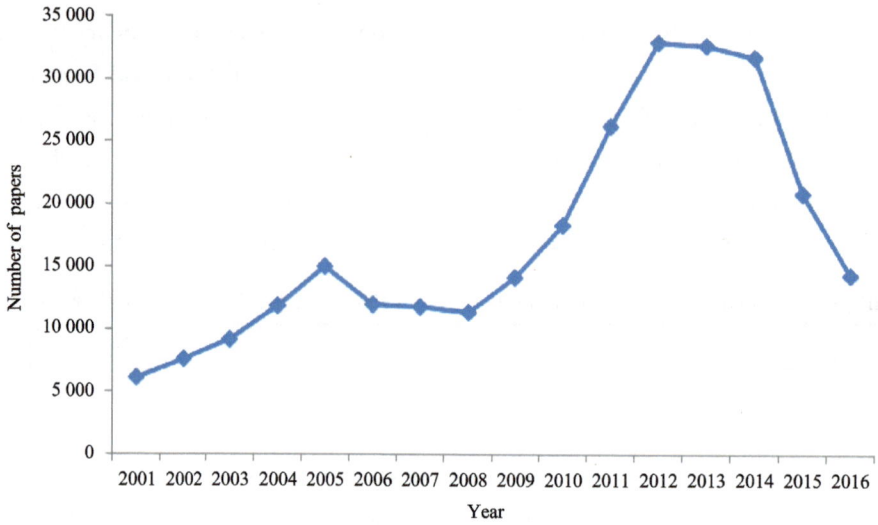

Figure 6-4　Annual Variation of the Number of EI Papers in Global Marine Field from 2001 to 2016

Figure 6-5 shows the number of EI papers published by the top 15 institutions in the global marine field. Chinese Academy of Sciences ranked first in the world. In addition, Harbin Engineering University, Dalian University of Technology and Ocean University of China were also among the top 15 institutions in the global marine field.

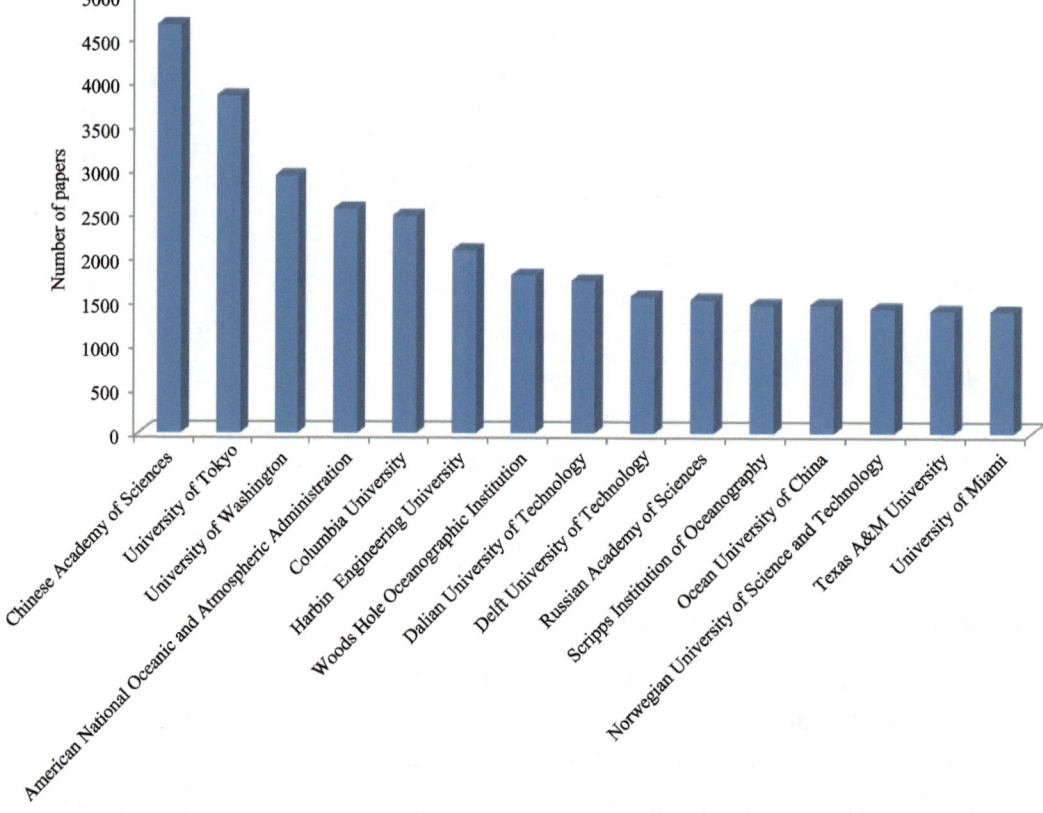

Figure 6-5　Top 15 Institutions with the Most EI Papers in Global Marine Field from 2001 to 2016

Figure 6-6 demonstrates 15 marine-related categories with the largest number of EI papers in marine field, which are mainly distributed in marine science and oceanography, seawater, tides and waves, general theory of oceanography, naval vessels among other fields. From the perspective of disciplinary distribution, a great deal of research is related to mathematics, materials science, mechanics, chemistry, biological engineering and biology, and computer application etc.

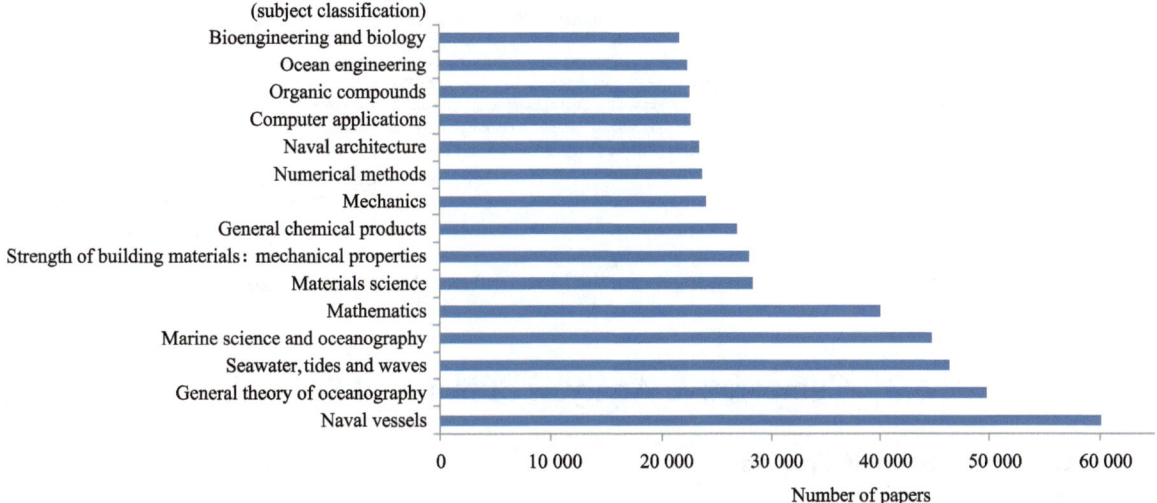

Figure 6-6　Fifteen Marine-related Fields with Most EI Papers from 2001 to 2016

Journal distribution of marine-related EI papers is very extensive. Figure 6-7 shows the 15 journals with the most published EI papers, which accounted for 9.94% of the total number of marine-related EI papers.

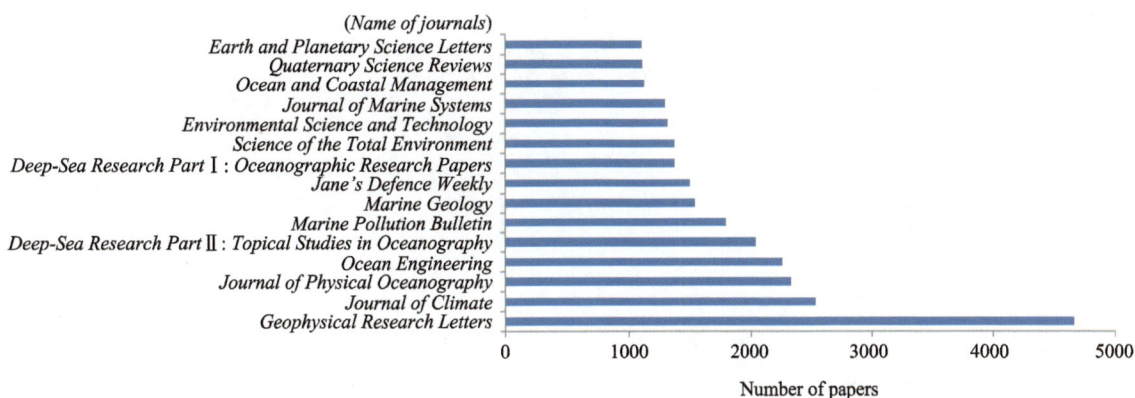

Figure 6-7　Number of Papers Published by Journals with the Most Marine-related EI Papers from 2001 to 2016

Conferences and EI papers from conferences are important channels to understand the marine-related research progress at home and abroad. Figure 6-8 demonstrates 15 conference proceedings which collected the largest number of marine-related papers. The major international conferences on the theme of marine research are International Offshore and Polar Engineering Conference, International Conference on Ocean, Offshore and Arctic Engineering, International Conference on Port and Ocean Engineering under Arctic Conditions, and the ISOPE Ocean Mining Symposium. In addition, there are also some national and regional conferences, such as Annual Offshore Technology Conference, the Coastal Engineering

Conference among others.

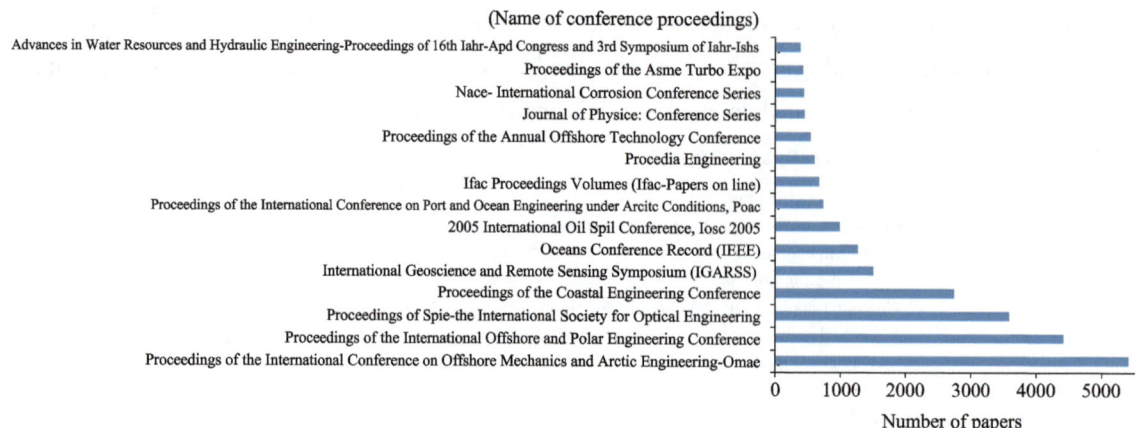

Figure 6-8　Conference Proceedings with the Most EI Papers Published from 2001 to 2016

6.2　Comparative Analysis of National Strength

6.2.1　Analysis based on SCI papers

Figure 6-9 shows the top 15 countries in the world with the most published SCI papers in the marine field according to the WOS database statistics from 2001 to 2016. Among them, the United States took the first place, followed by China and the United Kingdom, with more than 9500 papers. The rest of the top 15 countries were Australia, France, Germany, Canada, Japan, Spain, Russia, Italy, Norway, Netherlands, India, and Republic of Korea respectively.

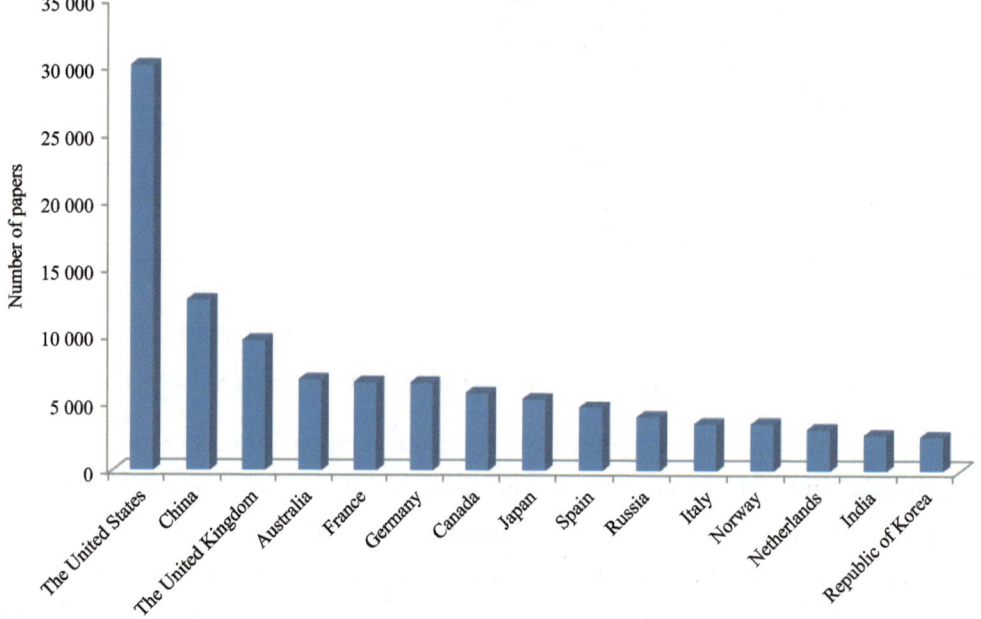

Figure 6-9　Number of SCI Papers in Global Marine Field Published by the Top 15 Countries from 2001 to 2016

Figure 6-10 shows the annual publications of SCI papers in the marine field of the top 15 countries from 2001 to 2016. The United States showed a steady growth trend. China witnessed a marked increasing trend, especially in the recent three years (2014-2016), while the United Kingdom, Australia, France, and Germany maintained relatively stable.

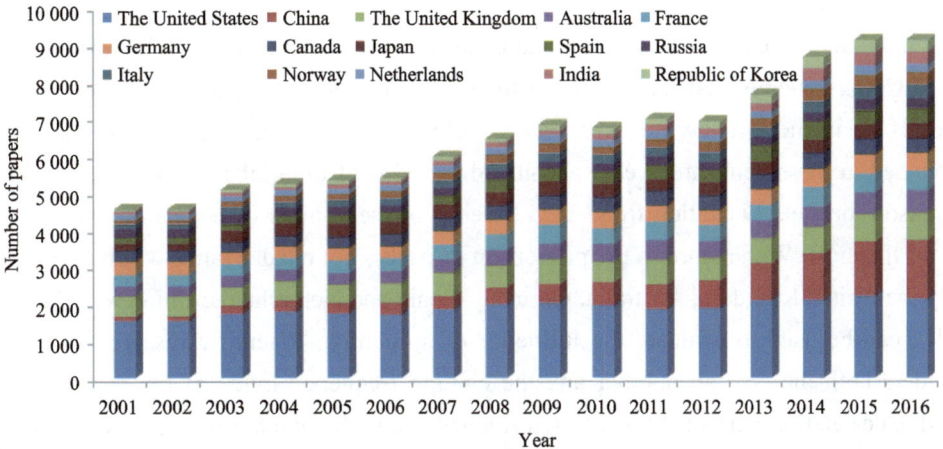

Figure 6-10   The Annual Publications of SCI Papers in the Marine Field of the Top 15 Countries from 2001 to 2016

Taking the WOS database as the data source, Table 6-2 calculated the research impact and output efficiency of the top 15 countries with the most published SCI papers in the global marine field, including the total citation frequency of papers, the average citation frequency of papers, the proportion of papers uncited and H index, and the number of papers published in the recent three years and its proportions.

Table 6-2   Statistics of the Impact and Output Efficiency of SCI Papers in Global Marine Field from 2001 to 2016

No.	Country	Number of papers	Average citation frequency /paper (times/paper)	Total citation frequency (times/paper)	Number of papers published in recent three years	Proportion of papers published in recent three years (%)	Number of papers uncited	Proportion of papers uncited(%)	H index
1	The United States	30 104	24.90	749 591	6 424	21	1 794	6	203
2	China	12 672	9.00	114 098	4 282	34	2 132	17	190
3	The United Kingdom	9 670	22.88	221 292	2 179	23	416	4	132
4	Australia	6 769	21.67	146 653	1 758	26	264	4	116
5	France	6 585	24.26	159 731	1 547	23	201	3	118
6	Germany	6 555	24.45	160 271	1 533	23	243	4	129
7	Canada	5 772	22.97	132 575	1 236	21	293	5	115
8	Japan	5 338	16.96	90 506	1 115	21	376	7	95
9	Spain	4 735	20.30	96 109	1 254	26	195	4	96
10	Russia	4 028	7.77	31 291	870	22	809	20	59
11	Italy	3 524	20.85	73 467	969	27	185	5	91
12	Norway	3 520	20.52	72 229	973	28	197	6	94
13	Netherlands	3 118	25.96	80 945	710	23	111	4	103
14	India	2 698	7.79	21 025	1 035	38	654	24	51
15	Republic of Korea	2 577	9.33	24 035	966	37	415	16	53

In terms of the research influence of major countries, the total citation frequency of papers from the United States ranked first, followed by the United Kingdom, Germany, France, Australia and Canada. Although the total number of Chinese papers published ranked second, the total citation frequency was in the seventh place. With regard to the average citation frequency of SCI papers in the marine field published by major countries, the Netherlands took first place (about 26 times/paper), and the United States, Germany and France all had more than 24 times/paper, and China had only 9 times/paper. The relatively low citation frequency of China generally can be attributed to the papers' lack of influence or the papers were only recently published in the last few years, resulting in hysteresis of influence. For example, Chinese SCI papers published in the recent three years accounted for 34% of the global total share, which may be an important reason for the low citation frequency. In terms of the number of papers that are not cited, China accounted for the most. With regard to the proportion of papers not cited, France was the least at about 3%, followed by the United Kingdom, Australia, Germany, Spain and Netherlands. China was 17%.

H index can be used to evaluate the influence of a country's scientific research papers, because H index considers both the citation number and the citation frequency index of the papers. H index has a strong positive correlation with the total citation frequency and the number of papers cited. The national H index can be explained as follows. Suppose Np papers are published in a country, if the citation number of H papers is higher than or equal to H, and the citation frequency of the other ($Np-H$) papers is lower than H, then the index value of scientific achievements of this country is H. Among the top 15 countries, the H indexes of the United States, China, the United Kingdom and Germany were relatively higher, indicating that these countries had outstanding scientific achievements in the marine field.

In terms of the output efficiency index of science and technology in major countries, the countries that published relatively more papers in the recent three years were the United States, China, the United Kingdom and Australia. In terms of the proportion of the papers published in the recent three years in all statistical years, China, India and Republic of Korea accounted for more than 30%, suggesting that marine innovation was on the rise in these countries.

6.2.2 Analysis based on EI papers

As can be seen from Figure 6-11, which shows the ten countries with most EI papers published in global

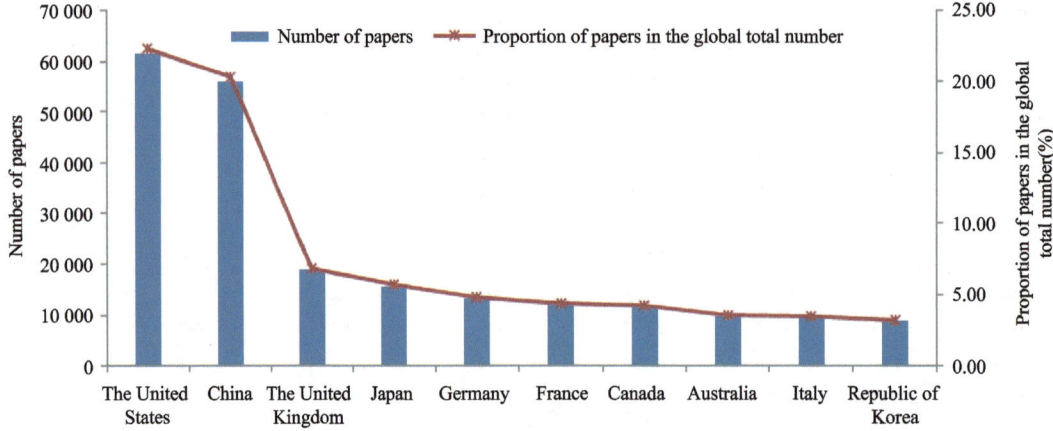

Figure 6-11 Top 10 Countries with the Most EI Papers in Global Marine Field and Their Proportions from 2001 to 2016

marine field from 2001 to 2016, the United States took first place, closely followed by China. The EI papers these two countries published accounted for about 40% of global publications and they were the two leading countries when it came to the production of EI papers in marine field.

Figure 6-12 shows the annual variation of the top ten countries with the largest number of EI papers published in the marine field. The annual growth rate of China and the United States in recent years was considerably higher than that of other countries. Since 2011, China has witnessed a particularly rapid increase in its number of papers published, which has even surpassed that of the United States and ranked first in the world.

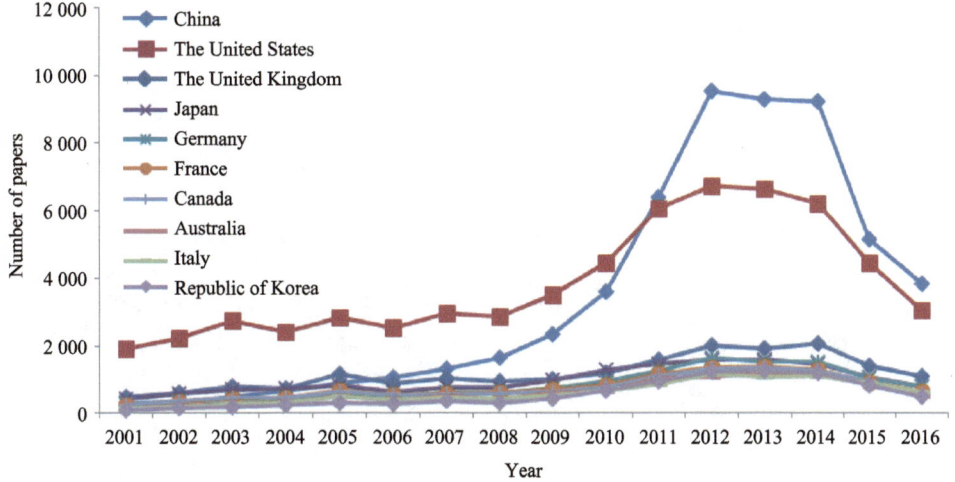

Figure 6-12　Top 10 Countries with Most EI Papers in the Marine Field and Their Annual Variation from 2001 to 2016

6.3　Analysis of Patented Technology Achievements in Marine Field

6.3.1　The overall research and development pattern in the world

On the basis of the data of marine patents from 2001 to 2016 retrieved from the DII database, as shown in Figure 6-13, the number of Chinese patent applications ranked the first, far ahead of other countries, regions and patent organizations. Chinese patent applications were equal to the total number of Republic of Korea, Japan, the United States and the International Bureau of the World Intellectual Property Organization (WIPO).

As shown in Figure 6-14, the increasing trend of marine patent applications in China was basically consistent with that of the entire world. China witnessed a steady increase in the share of the global marine patent applications, rising from 5.4% in 2001 to 84.4% in 2016 (including multi-national cooperation patent applications).

The number of marine patent applicantion in the world was increasing yearly, as shown in Figure 6-15, indicating that the marine industry was gradually growing stronger. The number of participants in the marine industry increased by nearly five times, rising from less than 1000 in 2001 to 5757 in 2015.

The number of global marine patent application institutions was also increasing yearly, as shown in Figure 6-16, from 949 in 2001 to 3 178 in 2016.

Figure 6-17 shows the main global marine patent application institutions. Five Chinese institutions entered the top 15, which were China National Offshore Oil Corporation, Zhejiang Ocean University, Ocean University of China, Zhejiang University, and Dalian Ocean University.

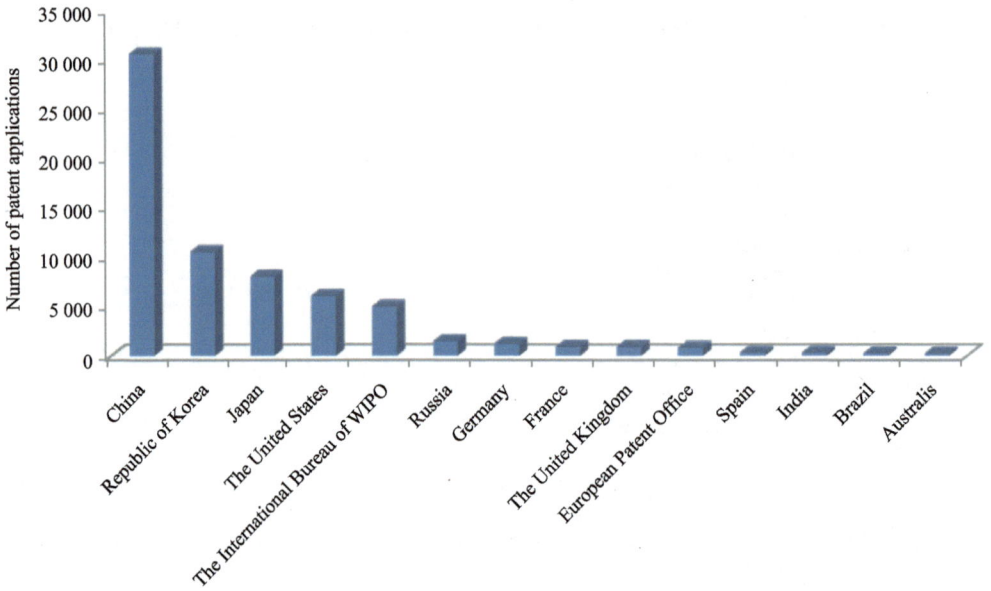

Figure 6-13　Distribution of Global Marine Patent Applications from 2001 to 2016

Taiwan Province of China is listed separately in order to more objectively reflect the situation of patent applications.

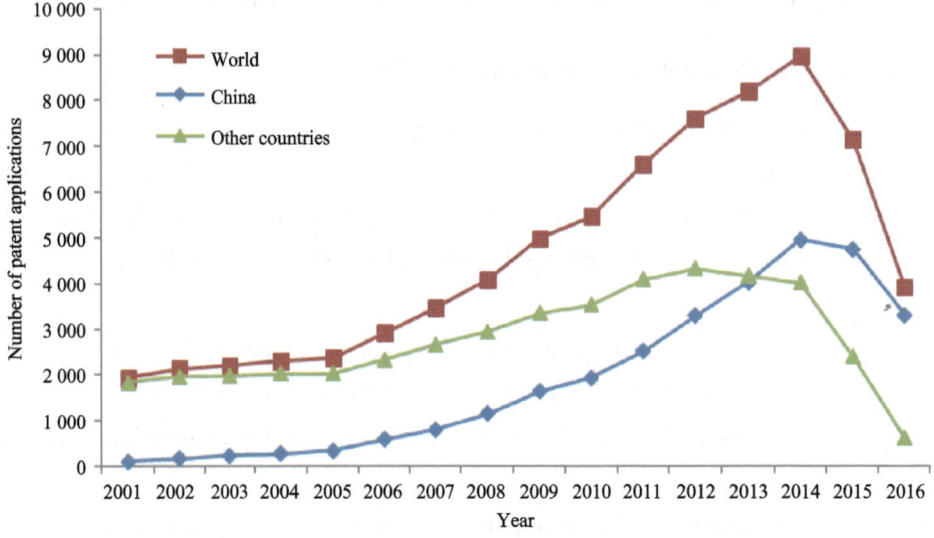

Figure 6-14　Annual Variation of the Number of Global Marine Patent Applications from 2001 to 2016

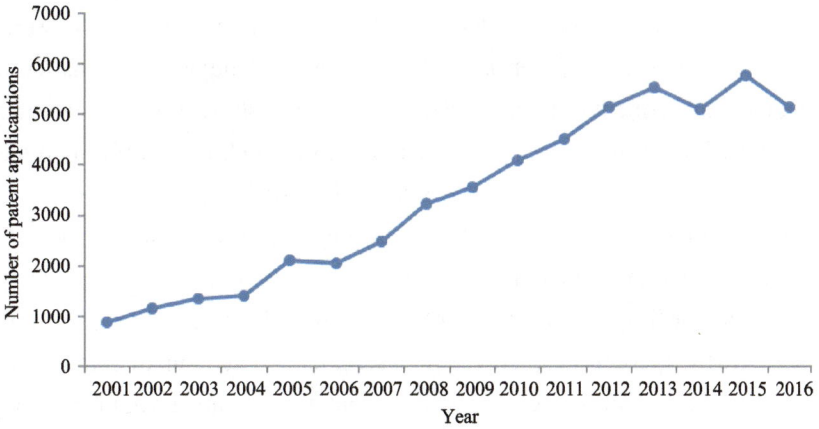

Figure 6-15   Annual Variation of the Number of Global Marine Patent Applicants from 2001 to 2016

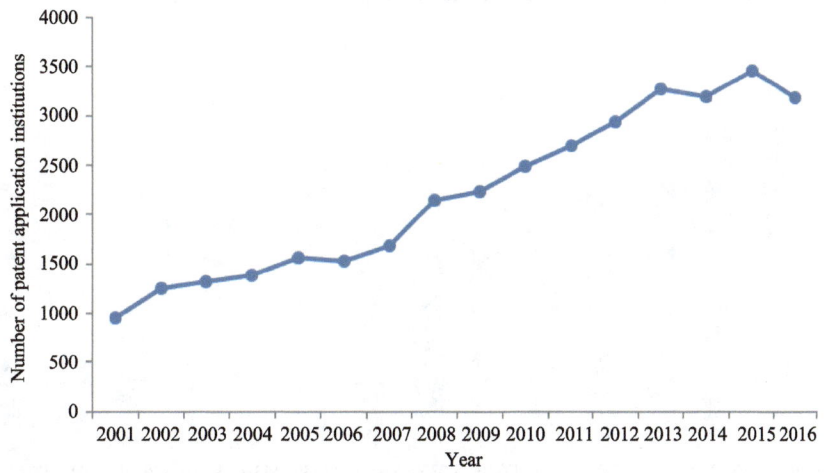

Figure 6-16   Annual Variation of the Number of Global Marine Patent Application Institutions from 2001 to 2016

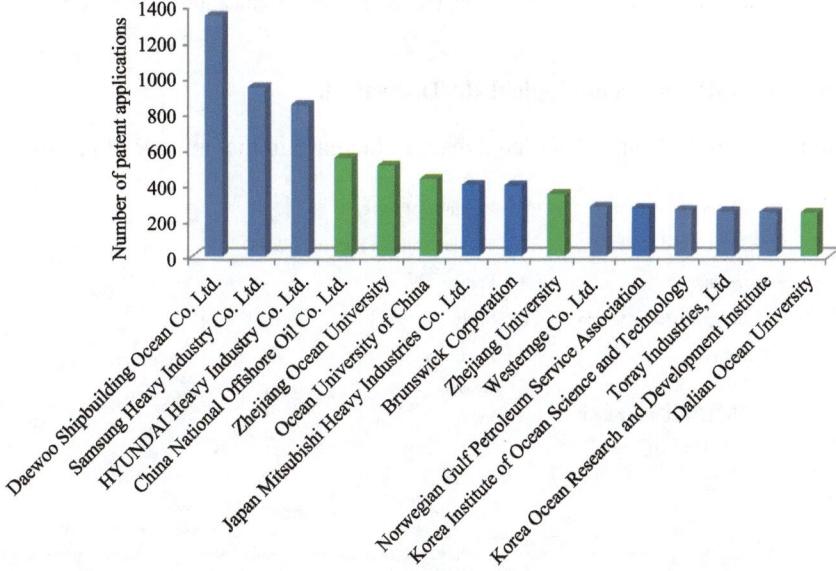

Figure 6-17   Comparison of the Top 15 Marine Patent Application Institutions in the Number of Global Marine Patent Applications from 2001 to 2016

As shown in Figure 6-18, the top 15 patents of global marine patents by IPC were B63B (ship and other waterborne vessels; marine equipment), C02F (sewage and sludge contamination treatment), A01K (animal husbandry; the management of poultry, fish, and insects; fishing; raising or breeding animals which are not included by other breeds; new breeds of animals), A23L (food, foodstuff or non-alcoholic beverages not included in A21D or subcategory from A23B to A23J), A61K(medical supplies), E02B (hydraulic engineering), F03B (hydraulic machinery or hydraulic engine), B01D (separation), A61P (therapeutic activity of compounds or pharmaceutical preparations), B63H (the propulsion device or steering device for ships), E21B (soil or rock drilling), G01N (testing or analyzing the material by measuring its chemical or physical properties), G01V (geophysics; gravity measurement; detection of matters or objects; tracer), E02D (foundation; excavation; filling), G01S (radio orientation; radio navigation; using radio wave for measuring distance or speed; measurement or speed; using radio wave reflection or the positioning or presence of re-radiation for detection; using similar devices of other waves).

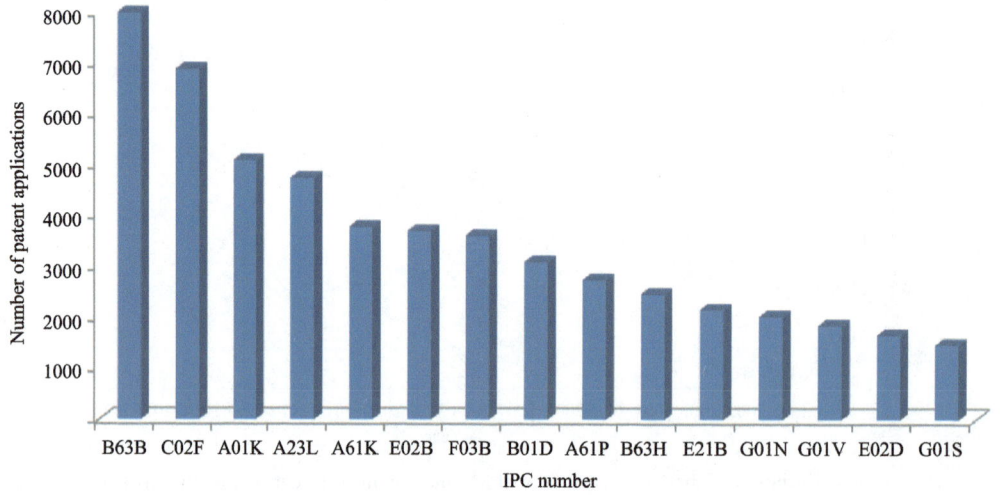

Figure 6-18　Comparison In the number of Global Marine Patent Applications by IPC from 2001 to 2016

6.3.2　Comparison of national technological R&D strength

As shown in Figure 6-19, China observed a marked increase in the number of patents according to the

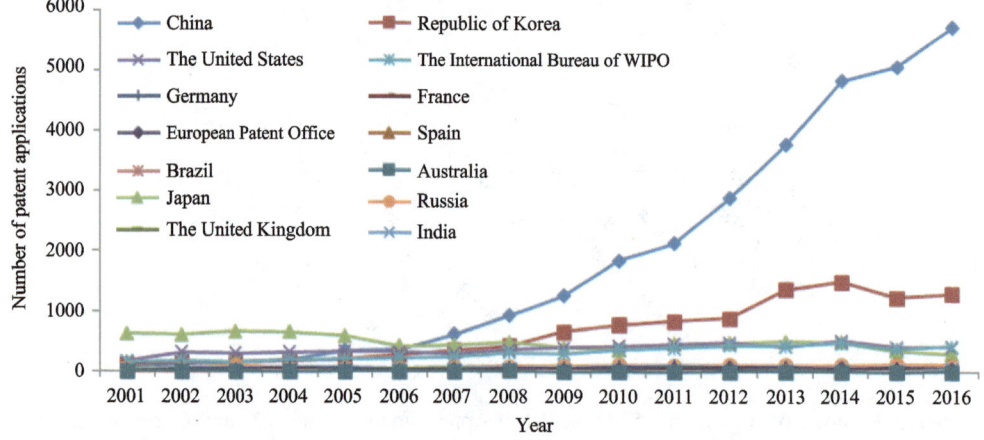

Figure 6-19　Annual Variation of the Number of Patent Applications (the Top 14 Countries) from 2001 to 2016

comparative analysis of the top 14 countries with the largest number of patent applications. Since the number of patent applications skyrocketed in 2006, China has taken the lead in the number of patent applications around the world, and the gap between the number of Chinese patent applications and that of other countries has widened. On the one hand, it indicates that China's marine industry has rapidly expanded in recent years; on the other hand, it may be related to the national patent supporting policies.

As can be seen from Figure 6-20, China accounted for more than 50% of the patent applications in the recent three years. Next were Republic of Korea and India which accounted for around 38%.

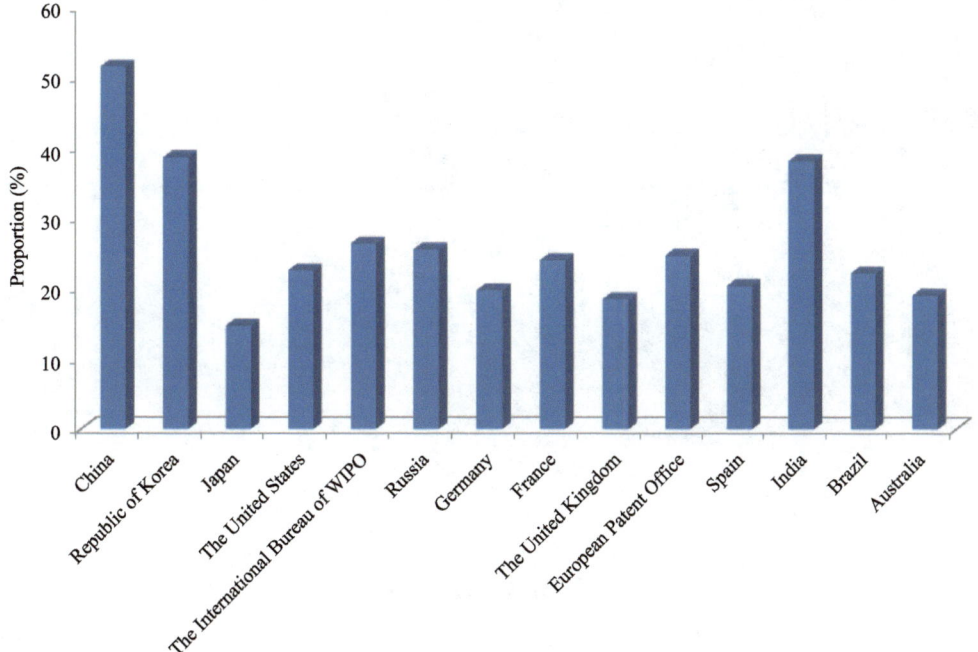

Figure 6-20   Proportion of Patent Applications in Recent Three Years (the Top 15 Countries) from 2001 to 2016

Appendixes

Appendix 1　Indicator System of National Marine Innovation Index

1. The Connotation of National Marine Innovation Index

National marine innovation index is a comprehensive index which measures the marine innovation capability and effectively reflects the marine innovation quality and efficiency of a country.

By using both Chinese and international theories and methodologies for evaluating a country's competitiveness and innovation, and based on the connotation analysis of innovative marine powers, we have identified the indicator selection principles and constructed an indicator system of national marine innovation index in four aspects: marine innovation resources, marine knowledge creation, marine innovation performance, and marine innovation environment. The indicator system is aimed at reflecting the characteristics of China's marine innovation capability at different levels of the innovation chain comprehensively, objectively and accurately, and forming a full-fledged indicator system and evaluation method. The index measurement can provide technical support and consulting service for comprehensively evaluating the construction progress of an innovative maritime power and improving marine innovation policies.

2. Connotation of Innovative Marine Power

To build a maritime power, there is urgent need to promote the shift of marine science and technology to an innovation-driven model. International historical experience shows that the development of marine science and technology is the fundamental guarantee for realizing a maritime power. Therefore, it is necessary to establish a national marine innovation evaluation index system to examine our country's marine development dynamics from a strategic perspective, strengthen marine basic research and the construction of talent teams, and vigorously develop marine science and technology to provide decision support for all sectors of economy and society.

The evaluation of national marine innovation index will help the national and local governments to keep abreast of the progress in the implementation of marine science and technology development strategies and possible problems, and provide basic information for further countermeasures. It is helpful for the international and domestic public to understand the progress, achievements, trends, and existing problems of China's marine undertakings. It will also help enterprises and investors to assess opportunities and risks in the marine field and provide relevant information for scholars and institutes engaged in marine research.

Throughout the history of marine economic development, China has undergone three stages: the stage of resource dependence, the stage of extensive expansion of industrial scale, and the stage of quantity-to-quality transition. With the rapid development of marine science and technology, the scale of new marine industries has been continuously expanding and became a new growth point for the marine economy. China has vast sea areas and abundant marine resources, but years of extensive development have made resources and environmental problems increasingly prominent, hindering further development of the marine economy. Thus, only by constantly carrying out marine innovation can we promote the healthy development of a marine economy and step into the rank of innovative maritime powers.

The most distinctive feature of an innovative maritime power is that fundamental changes have taken place in the mode of national marine economic and social development compared with the traditional mode of development. The judgement of whether a maritime power is innovation-oriented or not is based on what drives its marine economic growth. The marine economic growth of an innovative maritime power should be driven by innovative activities marked by knowledge creation, dissemination and application rather than by input of factors such as traditional consumption of marine resources and capital.

Innovative maritime powers should have the following four capabilities.

a. Higher comprehensive input capability of marine innovation resources.

b. Higher capabilities of marine knowledge creation, dissemination and application.

c. Better performance capabilities of marine knowledge creation.

d. Good environment for marine innovation.

3. The Selection Principles of Indicators

A. Evaluation reflects the idea of a sustainable marine development strategy

Consideration should be given not only to the holistic development environment of marine innovation but also the indicators such as sustainability of economic development and knowledge achievements, as well as the time trend of the indexes.

B. The data used in the report all come from authoritative sources

The basic data must come from national official statistics and surveys which are universally recognized. The statistics are collected through formal channels, on a timely and regular basis, to ensure accuracy, authoritativeness, continuity, and timeliness of basic data.

C. Indexes should be scientific, realistic and expandable

The marine innovation index has a close logical relationship with each sub-index. Every indicator of sub-index should be scientific, objective, and realistic to reduce the possibility of any artificial synthesis. Each indicator has a unique symbolic meaning at the macro level, defined in a relatively broad term rather than corresponding to the narrow scope of data, which facilitates the expansion and adjustment of the indicator system.

D. The evaluation system takes into account the characteristics of China's marine regions

Index selection is primarily based on relative index, with consideration given to the various characteristics of output efficiency of marine innovation resources, scale of innovation activities, and the width of innovation fields in different regions.

E. Historical analysis is combined with a country-by-country comparison

There are both retrospective analysis of historical development trends and comparisons between coastal areas, economic zones, economic circles, and a country-by-country analysis.

4. Construction of the Index System

Innovation is an entire process from the introduction of the innovation concept to research and development, from knowledge output to commercial applications, thus transforming knowledge into economic profits. The marine innovation capability is manifested in the whole process of creation, circulation, and transformation of marine S&T knowledge into economic profits. The indicator system should be constructed from the main links of the whole innovation chain, such as the marine innovation environment, the input of innovative resources, the creation and application of knowledge, and the impact of performance to evaluate national marine innovation capability.

The report adopts a comprehensive index evaluation method. Sub-indexes are selected from the perspective of the innovation process and four sub-indexes are identified, namely marine innovation resources, marine knowledge creation, marine innovation performance, and marine innovation environment are determined. Based on the principles of the indicator selection, twenty indicators are selected to form an index system for evaluating a national marine innovation index (Attached Table 1-1). All indicators are positive indicators. Then, comprehensive analysis, comparison, and assessment of the marine innovation capability of China based on national marine innovation comprehensive index and its indicator system have been conducted.

Attached Table 1-1　Indicator System of National Marine Innovation Index

Comprehensive index	Sub-index	Indicator	
National marine innovation index (A)	Marine innovation resources B_1	1. R&D funds input strength	C_1
		2. R&D personnel input strength	C_2
		3. Proportion of doctoral graduates in total R&D staff	C_3
		4. Proportion of S&T staff in total personnel of marine scientific research institutes	C_4
		5. Number of projects undertaken per 10 thousand scientific research personnel	C_5
	Marine knowledge creation B_2	6. Number of invention patent applications with economic output of 100 million US dollars	C_6
		7. Number of invention patent grants per 10 thousand R&D personnel	C_7
		8. S&T works published in the current year	C_8
		9. Number of S&T papers published per 10 thousand scientific research personnel	C_9
		10. Proportion of papers published abroad in total articles	C_{10}
	Marine innovation performance B_3	11. Transformation rate of marine S&T achievements	C_{11}
		12. Contribution rate of marine S&T progress	C_{12}
		13. Marine labor productivity	C_{13}
		14. Proportion of management services of scientific research education in GOP	C_{14}
		15. Marine economic output per unit energy of consumption	C_{15}
		16. Proportion of GOP in GDP	C_{16}
	Marine innovation environment B_4	17. Per capita GOP of coastal areas	C_{17}
		18. Proportion of equipment procurement cost in R&D funds	C_{18}
		19. Proportion of government funds in the total science and technology funds of marine scientific research institutes	C_{19}
		20. Full time equivalent workload per capita R&D personnel	C_{20}

Marine innovation resources reflect a country's investment in national marine innovation activities, the supply capacity of innovative talent resources and investment in infrastructure on which innovation relies.

Innovation input is a prerequisite for marine innovation activities of a country. This includes the funds' investment in science and technology and resources of talents.

Marine knowledge creation reflects a country's capabilities for marine scientific research output and knowledge dissemination. Marine knowledge creation is of various forms with multifaceted benefits. The report considers knowledge accumulation benefits of marine innovation from the perspectives of marine invention patents and scientific papers.

Marine innovation performance reflects the effect and impact brought about by a country's marine innovation activities. The marine innovation performance sub-index selects indicators from the efficiency and effectiveness of national marine innovation.

Marine innovation environment reflects the external environment on which a country's marine innovation activities depend, mainly including related marine system innovation and environment innovation. The main body of marine innovation system is the government and its related departments, which is manifested in the government's support for the policy related to innovation, and capital support to innovation, as well as the management of intellectual property rights. Environment innovation mainly refers to the allocation capability, infrastructure, fundamental economic level, finance, and cultural environment of innovation.

Appendix 2　Indicator Interpretations of National Marine Innovation Index

C_1. R&D Funds Input Strength

The indicator refers to the proportion of R&D funds of marine scientific research institutions in national GOP, namely the input strength indicator of national marine R&D funds, which reflects the strength of a country's investment in marine innovation.

C_2. R&D Personnel Input Strength

The indicator, which refers to the R&D personnel per 10 thousand marine employees, reflects the strength of a country's investment in human resources for innovation.

C_3. Proportion of Doctoral Graduates in Total R&D staff

The proportion of doctoral graduates in total R&D personnel reflects the top talents of a country's science and technology activities.

C_4. Proportion of S&T Staff in Total Personnel of Marine Scientific Research Institutes

The indicator, which refers to the proportion of S&T researchers in marine scientific research institutes, reflects the strength of scientific research power in marine innovation activities.

C_5. Number of Projects Undertaken per 10 Thousand Scientific Research Personnel

The indicator, which refers to the number of domestic projects undertaken per 10 thousand scientific researchers, reflects the intensity of marine scientific researchers engaged in innovative activities.

C_6. Number of Invention Patent Applications with Economic Output of 100 Million US Dollars

The indicator, which refers to the number of marine invention patent applications divided by GOP (in 100 million dollars converted from the exchange rate), reflects the level of technical output relative to economic output and the activity degree of a country's marine innovation activities. Invention patent has the highest technological content and value among the three types of patents (invention patent, utility model patent, and design patent). The number of invention patent applications can reflect the activity degree and independent innovation capability of country's marine innovation activities.

C_7. Number of Invention Patent Grants per 10 Thousand R&D Personnel

The indicator, which refers to the number of domestic invention patent grants per 10 thousand R&D personnel, reflects a country's independent innovation capability and technological innovation capability.

C_8. S&T Works Published in the Current Year

The indicator refers to scientific works, college textbooks and popular science books published by authorized publishing houses. Only books with the institution's scientific researchers as the first author are included. Books of the same title are counted as one work regardless of circulation. This indicator reflects a country's output capacity of scientific research.

C_9. Number of S&T Papers Published per 10 Thousand Scientific Research Personnel

The indicator reflects the output efficiency of scientific research.

C_{10}. Proportion of Papers Published Abroad in Total Number of Papers

The indicator, which refers to the proportion of papers published abroad in total S&T papers of a country, reflects the internationalization level of related research of scientific papers.

C_{11}. Transformation Rate of Marine S&T Achievements

The indicator, which measures the transformation of marine S&T innovation achievements into commercial products, refers to the proportion of a series of activities in total S&T achievements in order to increase level of productivity, which include a follow-up test, development, application, and promotion to transform marine S&T achievements with practical value as a result of scientific research and technological development into new products, new techniques, new materials and new industry.

C_{12}. Contribution Rate of Marine S&T Progress

The definition of this indicator is based on the definition of marine S&T growth rate. It refers to the proportion of the marine S&T growth rate in the total marine economic growth rate in various industries of marine economy. Marine S&T growth rate refers to the growth of other factors, when humans use marine resources and marine space for social production, exchange, allocation and consumption, but exclusive for production factors such as capital and labor. It specifically refers to the increase of the equipment technology level, the improvement of techniques, the enhancement of employee quality, and the strengthening of management and decision-making capability as a result of technological innovation, diffusion, transfer and introduction.

C_{13}. Marine Labor Productivity

The indicator, which refers to per capita GOP of employees engaged in marine-related jobs, reflects the role of marine innovation activities in marine economic output.

C_{14}. Proportion of Management Services of Scientific Research Education in GOP

The indicator reflects the contribution of marine scientific research, education, management, service and other activities to the marine economy.

C_{15}. Marine Economic Output per Unit Energy of Consumption

The indicator, which refers to GOP per 10 thousand standard-coal consumption, measures the reduction of energy consumption due to marine innovation and reflects a country's intensification level of marine economic growth.

C_{16}. Proportion of GOP in GDP

The indicator reflects the contribution of marine economy to national economy and measures the role of marine innovation in promoting marine economy.

C_{17}. Per Capita GOP of Coastal Areas

The indicator, to some degree, reflects the living standards in coastal areas, and measures the growth status of marine productivity and the external environment for marine innovation activities.

C_{18}. Proportion of Equipment Procurement Cost in R&D Funds

The indicator, which refers to the proportion of equipment procurement cost in R&D funds in marine scientific research institutes, shows the hardware requirements that marine innovation needs. This indicator reflects the hard environment of marine innovation to some degree.

C_{19}. Proportion of Government Funds in the Total Science and Technology Funds of Marine Scientific Research Institutes

The indicator reflects the role of government investment in promoting marine innovation and the system environment for marine innovation.

C_{20}. Full Time Equivalent Workload per Capita R&D Personnel

The indicator reflects the workload and full-time working capacity of a country's marine science and technology human resources.

Appendix 3　Evaluation Methods of National Marine Innovation Index

The calculation method of the national marine innovation index adopts the internationally popular benchmarking method, namely the method adopted by the Lausanne's International Competitiveness

Evaluation. Benchmarking is a methodology widely used in the world today. It works in the following way: first, a benchmark value is set, then the subjects of evaluation will be measured against the benchmark value to identify the gap between each other and finally show the results of ranking.

Based on the indicators of the evaluation indicator system of marine innovation and the indicator data from 2002 to 2016, the scores of the marine innovation index and sub-indexes in years afterwards can be calculated separately. The growth of the national marine innovation index can be seen by comparison with the base year.

1. Standardized Processing of the Original Data

With 2004 as the base year and 100 as the base value, the original value of the 20 indicators in the indicator system of marine innovation index was standardized as follows:

$$C_j^t = \frac{100 x_j^t}{x_j^1}$$

In the formula, j is the sequence number of the indicator, $j=1$-20; t is the number of years 2004-2016, $t=1$-13; x_j^t is the original data for the reference year's indicators (x_j^1 is the original data of indicators of 2004); C_j^t is the value after standardized processing of the indicators.

2. Calculation of National Marine Innovation Sub-index

The scores for the sub-indexes are calculated based on equal weight[①] (the same as below):

When $i=1$, $B_1^t = \sum_{j=1}^{5} \beta_1 C_j^t$, $\beta_1 = \frac{1}{5}$;

When $i=2$, $B_2^t = \sum_{j=6}^{10} \beta_2 C_j^t$, $\beta_2 = \frac{1}{5}$;

When $i=3$, $B_3^t = \sum_{j=11}^{16} \beta_3 C_j^t$, $\beta_3 = \frac{1}{6}$;

When $i=4$, $B_4^t = \sum_{j=17}^{20} \beta_4 C_j^t$, $\beta_4 = \frac{1}{4}$;

In the formula, β is the weight, $t=1$-13, B_1^t, B_2^t, B_3^t, B_4^t respectively represents the scores for the marine innovation resources sub-index, marine knowledge creation sub-index, marine innovation performance sub-index, and marine innovation environment sub-index.

3. Calculation of National Marine Innovation Index

The scores of national marine innovation index are calculated based on equal weight (ibid):

$$A^t = \sum_{i=1}^{5} \varpi B_i^t$$

In the formula, $i=1$-4; $t=1$-13; ϖ is the weight (equal weight = $\frac{1}{4}$); A^t is the score of national marine innovation index in different years.

---

① Equal weight is obtained based on weight selection method of the *National Innovation Index Report 2016*

Appendix 4 Evaluation Methods of Regional Marine Innovation Index

1. Explanations on the Indicator System of Regional Marine Innovation Index

The indicator system of regional marine innovation index is composed of four sub-indexes of marine innovation resources, marine knowledge creation, marine innovation performance, and marine innovation environment. Compared with the sub-index of national marine innovation performance, the sub-index of regional marine innovation performance lacks two indicators of the "contribution rate of marine S&T progress" and "transformation rate of marine S&T achievements".

2. Normalization of Raw Data

The original values of the 18 indicators in 2016 are normalized respectively. Normalization processing is used for the purpose of removing the discrepancies of the measurement unit, the differences of order of magnitude and relative number of indicator values when doing a multi-indicator comprehensive evaluation, thereby solving the comparability issue of the data indicator and putting indicators in the same magnitude order for the convenience of comprehensive analysis and contrast.

The indicator data is processed based on linear normalization:

$$c_{ij} = \frac{y_{ij} - \min y_{ij}}{\max y_{ij} - \min y_{ij}}$$

In the formula, i is the sequence number of eleven coastal provinces (autonomous regions, municipalities directly under the central government) in Chinese mainland, i=1-11; j is the sequence number of the indicator, j =1-18; y_{ij} is the original data value of the indicators; c_{ij} is the normalized data value.

3. Calculation of the Sub-indexes of Regional Marine Innovation

The score of the sub-indexes of regional marine innovation environment:

$$b_1 = 100 \times \sum_{j=1}^{5} \varphi_1 c_j, \quad \varphi_1 = \frac{1}{5}$$

The score of sub-indexes of regional marine knowledge creation:

$$b_2 = 100 \times \sum_{j=6}^{10} \varphi_2 c_j, \quad \varphi_2 = \frac{1}{5}$$

The score of sub- indexes of regional marine innovation performance:

$$b_3 = 100 \times \sum_{j=11}^{14} \varphi_3 c_j, \quad \varphi_3 = \frac{1}{4}$$

The score of sub-indexes of regional marine innovation environment:

$$b_4 = 100 \times \sum_{j=15}^{18} \varphi_4 c_j, \quad \varphi_4 = \frac{1}{4}$$

In the formula, $j=1\text{-}18$, b_1, b_2, b_3, b_4 respectively represent the sub-index scores: regional marine innovation resources, regional marine knowledge creation, regional marine innovation performance, and regional marine innovation environment.

4. Calculation of Regional Marine Innovation Index

The scores of regional marine innovation index are calculated based on equal weight (the same as the national marine innovation index):

$$a = \frac{1}{4}(b_1 + b_2 + b_3 + b_4)$$

In the formula, a is the score of the regional marine innovation index.

Appendix 5 Calculation Methods of Contribution Rate of Marine S&T Progress

Currently, the widely used method for calculating the contribution rate of S&T progress is the Solow Residual Method. It is also a method commonly used by National Development and Reform Commission (NDRC, formerly the National Planning Commission), the National Bureau of Statistics and the Ministry of Science and Technology, etc.

With Cobb-Douglas Production Function as a base model, Solow Residual Method indicates that economic growth depends not only on the capital growth rate, labor growth rate and the weight of the relative effect that capital and labor have on income growth, but also on technological progress, thus distinguishing "growth effect" caused by the increase of quantities of different factors from the "level effect" of economic growth as a result of the improvement of technological levels, and systematically explaining the reasons for economic growth.

The marine economy involves many industries and sectors. In order to comprehensively reflect the contribution of S&T progress in various marine industries to the overall increase of the marine economy, calculation and measurement across the board need to be conducted to all types of marine industries. Then, according to the proportion of total economic output of various industries in the marine economy and by aggregating and weighting marine S&T progress of various industries at the measurement stage of growth rate, we obtained the growth rate of marine S&T progress and then the contribution rate of marine S&T progress after further calculation.

The industries and sectors that the contribution rate of marine S&T progress involves based on its theoretical connotation and characteristics are as follows: production and service of products obtained directly from the ocean; one-time processing production and service of products obtained directly from the ocean; production and service of products which are directly applied to the ocean; production and service using seawater and ocean space as the basic elements of production process. Other services and management scopes such as marine scientific research, education and technology are not appropriate to be calculated by the contribution rate of marine S&T progress.

In combination with our country's marine S&T characteristics and through industrial weighting of the output growth rate, capital growth rate, and labor growth rate of eight marine industries, the basic formula to calculate the contribution rate of marine S&T progress is constructed as follows.

With "i" industry (i = 1, 2, 3, ⋯, 8) representing marine culture, marine fishing, marine salt industry, marine shipping, marine oil, marine natural gas, marine transportation, and coastal tourism respectively; $y_i(t)$ indicates the output growth rate of the ith industry during t period, among which $t \in [t_1, t_2]$; $k_i(t)$ and $l_i(t)$ indicate the growth rate of capital and labor input within t period respectively, among which $t \in [t_1, t_2]$; γ_i indicates the weight of the ith industry in overall marine industries; k_i, l_i, y_i indicate the average value of $k_i(t), l_i(t), y_i(t)$ respectively between t_1 and t_2, namely:

$$k_i = \frac{\sum_{t=t_1}^{t_2} k_i(t)}{n}, \quad l_i = \frac{\sum_{t=t_1}^{t_2} l_i(t)}{n}, \quad y_i = \frac{\sum_{t=t_1}^{t_2} y_i(t)}{n}, \quad (n = t_2 - t_1)$$

$$k = \sum_{i=1}^{8} k_i \gamma_i, \quad l = \sum_{i=1}^{8} l_i \gamma_i, \quad y = \sum_{i=1}^{8} y_i \gamma_i$$

In the formula, k, l, y indicate weighted average of k_i, l_i, y_i.

Thus leading to the following formula:

$$A = 1 - \frac{\alpha k}{y} - \frac{\beta l}{y} = 1 - \frac{\alpha \sum_{i=1}^{8} k_i \gamma_i}{\sum_{i=1}^{8} y_i \gamma_i} - \frac{\beta \sum_{i=1}^{8} l_i \gamma_i}{\sum_{i=1}^{8} y_i \gamma_i}$$

$$= 1 - \frac{\alpha \sum_{i=1}^{8} \frac{\sum_{t=t_1}^{t_2} k_i(t)}{n} \gamma_i}{\sum_{i=1}^{8} \frac{\sum_{t=t_1}^{t_2} y_i(t)}{n} \gamma_i} - \frac{\beta \sum_{i=1}^{8} \frac{\sum_{t=t_1}^{t_2} l_i(t)}{n} \gamma_i}{\sum_{i=1}^{8} \frac{\sum_{t=t_1}^{t_2} y_i(t)}{n} \gamma_i}$$

In the formula, A is the contribution rate of marine S&T progress within the period of research; α and β indicate marine industry capital and elasticity coefficient of labor respectively.

In terms of the indicator length selection, measurement time length used to calculate the contribution rate of national marine S&T progress should be more than ten years, or at least five years, because the impact of marine S&T on the marine economy is long-term. Considering the actual need of marine management and time limit of marine data, the study uses the average value of five-year data when calculating the indicators of the 11th Five-Year Plan period and doing a short-term forecast to the indicators of the 12th Five-Year Plan. The average value of the ten-year data is used when doing other calculations and a long-term forecast (depending on the time length from 2006 to 2016).

In terms of marine industry selection, according to *China Marine Statistical Yearbook 2017*, the twelve major marine industries in 2016 were as follows: marine fishery (16.26%), marine oil and gas industry (3.06%), marine mining (0.24%), marine salt industry (0.14%), marine shipping industry (5.26%), marine chemistry industry (3.39%), marine bio pharmaceutics industry (1.20%), marine engineering construction industry (6.10%), marine electric power industry (0.45%), seawater utilization industry (0.05%), marine transportation industry (20.08%), and coastal tourism industry (43.79%) (Attached Table 5-1). After a preliminary screening and feasibility analysis, eight industries can be calculated and measured with supporting data, including marine culture, marine fishing, marine salt, marine shipping, marine oil, marine natural gas, marine transportation, and coastal tourism. The total production value of the above mentioned eight industries accounted for 86.96% of the total production value of the major marine industries, which can effectively reflect the development of China's marine economy.

Attached Table 5-1 Added Value of the Major Marine Industries of China in 2016

Major marine industry	Added value (hundred million yuan)	Proportion (%)
Marine fishery	4 615.4	16.26
Marine oil and gas industry	868.8	3.06
Marine mining	67.3	0.24
Marine salt industry	38.9	0.14
Marine shipping industry	1 492.4	5.26
Marine chemistry industry	961.8	3.39
Marine bio pharmaceutics industry	341.3	1.20
Marine engineering construction industry	1 731.3	6.10
Marine electric power industry	128.5	0.45
Seawater utilization industry	13.7	0.05
Marine transportation industry	5 699.8	20.08
Coastal tourism industry	12 432.8	43.79
Total	28 392	—

In terms of the determination of elasticity coefficient, the elasticity coefficient of capital and labor output can be determined by adopting the Experience Estimation Method, Ratio Method and Regression Method when calculating the contribution rate of marine S&T progress. Experience Estimation Method means using the coefficient calculated by other authoritative experts for reference. The principle of Ratio Method is to make use of the date related to the amount of capital input and labor input to calculate their ratio. Regression Method refers to the use of constrained production function model (namely $\alpha+\beta=1$) or unconstrained production function model to estimate two coefficients after locating relevant numerical values according to the measurement method (namely using the Least Square Method to regress). The method adopted by the study is: $\alpha=0.3$, $\beta=0.7$.

In terms of the determination of weightings, the weight values (Attached Table 5-2) of various industries are determined according to the output value of the eight industries during the 12th Five-Year Plan period mentioned in *China Marine Statistical Yearbook*.

Attached Table 5-2 Weight Values of Each Industry

Industry	Weight value	Industry	Weight value
Marine culture	0.1096	Marine oil	0.0709
Marine fishing	0.0810	Marine natural gas	0.0045
Marine salt	0.0003	Marine transportation	0.2489
Marine shipping	0.0664	Coastal tourism	0.4154

With regard to data sources, all the indicator data representing marine industry value, capital and labor adopted by this study are from *China Marine Statistical Yearbook* in the corresponding year (Attached Table 5-3). From the data base, the continuous data that can be calculated are the output value of marine industries, capital and labor from 1996 to 2016 (trend fitting interpolation are used to deal with individual missing data).

Attached Table 5-3 Indicators of the Output, Capital and Labor of the Eight Industries

Industry	Output indicator	Capital indicator	Labor indicator
Marine culture	Output of marine culture	Area of marine culture	Number of the employed of marine fishing and relevant industries
Marine fishing	Output of marine fishing	Total tons of ships working on the sea	Number of the employed of marine fishing and relevant industries
Marine salt	Sea salt production of coastal areas	Production area of marine salt	Number of the employed of marine salt industry
Marine shipping	Added value of marine shipping	Completion amounts of ship building of coastal areas	Number of the employed of marine shipping
Marine oil	Marine crude oil production of coastal areas	Offshore oil wells	Number of the employed of marine oil and natural gas industry
Marine natural gas	Marine natural gas production of coastal areas	Offshore gas wells	Number of the employed of marine oil and natural gas industry
Marine transportation	Added value of marine transportation	Number of berths used for production in ports above designated size of coastal areas	Number of the employed of marine transportation industry
Coastal tourism	Added value of coastal tourism	Total number of travel agencies of coastal areas	Number of the employed of coastal tourism

The average value of the contribution rate of marine S&T progress can be obtained by locating the benchmark data of each industry within the formula measuring the contribution rate of marine S&T progress. After calibration and verification, the average values of the contribution rate of marine S&T progress in China during the 11th Five-Year Plan period and the period from 2006 and 2016 were 54.4% and 65.9% respectively.

Appendix 6 Calculation Methods of Transformation Rate of Marine S&T Achievements

The definition for the transformation rate of marine S&T achievements is derived from S&T achievements transformation rate. With regard to the studies on the transformation rate of S&T achievements, foreign scholars rarely use the term "S&T achievements transformation" directly, but tend to replace it with "integration of technology and economy" "technology innovation" "technology transformation" "technology promotion" "technology diffusion" or "technology transfer". In addition, there is no statistical analysis or evaluation done to transformation situations of S&T achievement of the whole society in foreign countries.

Domestically, scholars in various fields are divided in the definitions of S&T achievements transformation, which can be reduced to the following three viewpoints.

Viewpoint 1: The transformation rate of S&T achievements refers to the ratio of transformed S&T

achievements in all applied technological achievements. Scholars hold that "the transformed S&T achievements" do not refer to all S&T achievements that have been "transformed". We should investigate the acceptance of the market to the technological achievements or their direct or indirect benefits when technology achievements are applied to production. If the applied technology achievements can be successfully transformed into commodities and achieve economies of scale, then the applied technology achievements have realized transformation.

Viewpoint 2: The transformation rate of S&T achievements refers to the ratio of transformed S&T achievements in all S&T achievements. Scholars hold that although most of the fundamental theoretical achievements and some soft science achievements cannot be directly applied to the actual production and the quantitative degree of achievement transformation is low, they are still able to promote the progress of science and technology and the adjustment and optimization of industrial structure to a certain extent. Therefore, it is recommended that the transformation of basic theoretical achievements and soft science achievements should be incorporated into S&T achievement transformation.

Viewpoint 3: From the perspective of management, the S&T achievement transformation rate should show the ratio of S&T achievement in all research subjects.

The second viewpoint should not be adopted because the fundamental research achievements and soft science research achievements in the marine field can hardly be directly applied to the actual production, making it difficult to realize the transformation of S&T achievements. With regards to the third viewpoint, "S&T achievements" and "research subjects" that the definitions involve come from two sets of different marine statistical data with the former one deriving from the marine S&T statistical data while the latter one from the marine S&T achievements statistical data, so this viewpoint cannot accurately reflect the actual transformation of marine S&T achievements.

Therefore, this report adopts the first viewpoint with the definition of transformation rate of marine S&T achievements as follows.

The transformation rate of marine S&T achievements refers to the percentage of marine S&T achievements of marine-related enterprises within a certain period in the total application of marine S&T achievements. These marine S&T achievements can realize self-transformation or be transformed into production, and they must be at the application or production stage and have reached maturity in application. According to the definition of marine S&T achievement transformation rate, the formula for marine S&T achievement transformation rate can be constructed as follows.

Transformation rate of marine S&T achievements= maturely applied marine S&T achievements / all achievements of marine S&T applied technology × 100%

As the transformation of marine S&T achievements is a long-term process, the longer the coverage period is, the more realistic the indicators are when measuring and calculating the transformation rate of marine S&T achievements.

It should be noted that the transformation rate of marine S&T achievements discussed in this report refers to indicators in a narrow sense. The "maturely applied marine S&T achievement" and "all achievements of marine S&T applied technology" in the formula come from registered data of marine S&T

achievements. In a broad sense, marine scientific research projects, patents, papers, awards, standards, software and copyright all belong to marine S&T achievements, therefore it is difficult to obtain the statistics and there is overlapping between each other. It is extremely hard to identify and measure the process because all steps from the formation of marine S&T achievements to the initial application and then to the formation of products, until achieving scale and industrialization, can be counted into marine S&T achievements transformation process.

Based on the statistical data of marine S&T achievements, the transformation rate of marine S&T achievements in our country in 2016 was approximately 50.0% according to the calculation using the standard formula for the transformation rate of marine S&T achievements.

According to the S&T achievements registration form, the applied technology achievements can be divided into three stages. The initial stage refers to the research achievements of laboratory (small test) at the early stage. Mid-term stage refers to the intermediate test (pilot test) for further improvement of the product, technique or production process before the new products, new technique and new production process are directly used for production; the prototype and the sample made for finalizing product design and obtaining the technical parameters needed for production; demonstration for wide promotion; phased research achievements for mature application and wide promotion. Mature application stage refers to the achievements that have realized industrialization and been officially (or can be officially) applied, including large-scale promotion of agricultural technology, clinical application of medical health care and other achievements such as the prototype and final product design of public security and military industry.

Appendix 7 Regional Classification Basis and Definition of Related Concepts

1. Coastal Provinces (Autonomous Regions, Municipalities Directly under the Central Government)

Eleven provinces (autonomous regions, including municipalities directly under the central government) have coastlines, including Tianjin, Hebei, Liaoning, Shanghai, Jiangsu, Zhejiang, Fujian, Shandong, Guangdong, Guangxi and Hainan.

2. Marine Economic Zones

China has five marine economic zones, namely, the Bohai Rim Economic Zone, the Yangtze River Delta Economic Zone, the West Coast of the Taiwan Straits Economic Zone, the Pearl River Delta Economic Zone, and the Beibu Gulf Rim Economic Zone. Coastal provinces (including municipalities directly under the central government) which are incorporated in the evaluation of the Bohai Rim Economic Zone are Liaoning, Hebei, Shandong and Tianjin; coastal provinces (municipalities directly under the central government) which are incorporated in the evaluation of the Yangtze River Delta Economic Zone are Jiangsu, Shanghai and Zhejiang; Fujian is incorporated in the evaluation of the West Coast of the Taiwan Straits Economic Zone; Guangdong is incorporated in the evaluation of the Pearl River Delta Economic Zone; and the coastal provinces (autonomous regions) which are incorporated in the evaluation of the Beibu Gulf Rim Economic Zone are Guangxi and Hainan.

3. Marine Economic Circles

China boasts three marine economic circles including northern marine economic circle, eastern marine economic circle, and southern marine economic circle based on the *The 12th Five-Year Plan for National Marine Economic Development*. The northern marine economic circle comprises Liaodong Peninsula, the Bohai Bay and the coasts and sea areas of Shandong Peninsula and the coastal provinces (municipalities directly under the central government) which are incorporated in this evaluation include Tianjin, Hebei, Liaoning and Shandong; the eastern marine economic circle is composed of Jiangsu, Shanghai and the coasts and sea areas of Zhejiang, and the coastal provinces (municipalities directly under the central government) which are incorporated in this evaluation include Jiangsu, Zhejiang and Shanghai; the southern marine economic circle comprises Fujian, the Pearl River Estuary and its two wings, the Beibu Gulf and the coasts and sea areas of Hainan Island, and the coastal provinces (autonomous regions) which are incorporated in this evaluation include Fujian, Guangdong, Guangxi and Hainan.

Appendix 8 List of Marine-related Higher Institutions (Including Marine-related Coefficient of Proportionality)

1. Higher Institutions Directly under the Ministry of Education

Peking University (0.0932) (the marine-related coefficient of proportionality is determined according to the proportion of marine-related majors in the total number of majors at Peking University; the same below), Tsinghua University (0.0256), Beijing Normal University (0.1373), China University of Geosciences (Beijing) (0.2381), Tianjin University (0.0877), Dalian University of Technology (0.0886), Shanghai Jiao Tong University (0.0484), Nanjing University (0.1163), Hohai University (0.9020), Zhejiang University (0.1102), Xiamen University (0.0707), Ocean University of China (0.8462), Wuhan University (0.0645), China University of Geosciences (Wuhan) (0.2258), Sun Yat-sen University (0.1280), Tongji University (0.0859), East China Normal University (0.0789), Huazhong University of Science and Technology (0.0566), South China University of Technology (0.0490).

2. Higher Institutions Directly under the Ministry of Industry and Information Technology

Harbin Institute of Technology (0.0462).

3. Higher Institutions Directly under the Ministry of Transport

Dalian Maritime University (0.9348).

4. Local Higher Institutions

Shanghai Ocean University (0.3191), Guangdong Ocean University (0.2200), Dalian Ocean University (0.9545), Zhejiang Ocean University (0.8913), Ningbo University (0.1935), Jimei University (0.2388), Nanjing University of Information Science & Technology (0.2759), Hainan Tropical Ocean University (0.1964).

Appendix 9 List of Marine-related Disciplines (Discipline Classifications of the Ministry of Education)

Attached Table 9-1 List of Marine-related Disciplines (Discipline Classifications of the Ministry of Education)

Code	Disciplines	Descriptions
140	Physics	
14020	Acoustics	
1402050	Underwater Acoustics and Ocean Acoustics	Formerly known as "Hydroacoustics".
14030	Optics	
1403064	Ocean Optics	
170	Geoscience	
17050	Geology	
1705077	Geology of Petroleum and Natural Gas	Including Geology of Natural Gas Hydrate.
17060	Marine Science	
1706010	Marine Physics	
1706015	Marine Chemistry	
1706020	Marine Geophysics	
1706025	Marine Meteorology	
1706030	Marine Geology	
1706035	Physical Oceanography	
1706040	Marine Biology	
1706045	Marine Geography & Estuarine and Coastal Science	Formerly known as "Estuarine and Coastal Science".
1706050	Marine Investigation and Monitoring	
	Marine Engineering	See 41630.
	Marine Surveying and Charting	See 42050.
1706061	Remote Sensing Oceanography	Also called "Satellite Oceanography".
1706065	Marine Ecology	
1706070	Environmental Oceanography	
1706075	Science of Marine Resources	
1706080	Polar Science	
1706099	Other Disciplines of Marine Science	
240	Fishery Science	
24010	Basic Disciplines of Fishery Science	
2401010	Aquatic Chemistry	
2401020	Geography of Fishery	
2401030	Aquatic Biology	
2401033	Aquatic Genetics and Breeding Science	
2401036	Aquatic Animal Medicine Science	
2401040	Aquatic Ecology	

Appendixes

Continued

Code	Disciplines	Descriptions
2401099	Other Disciplines of the Basic Disciplines of Fishery Science	
24015	Aquatic Multiplication Science	
24020	Aquaculture science	
24025	Aquatic Feed Science	
24030	Aquatic Protection Science	
24035	Piscatology	
24040	Storage and Processing of Aquatic Products	
24045	Aquatic Engineering	
24050	Aquatic Resource Science	
24055	Aquatic Economics	
24099	Other Disciplines of Fishery Science	
340	Military Medicine and Special Medicine	
34020	Special Medicine	
3402020	Diving Medicine	
3402030	Nautical Medicine	
413	Engineering and Technology Related to Information and System Science	
41330	Systemic Application of Information Technology	
4133030	Marine Information Technology	
416	Engineering and Technology Related to Natural Science	
41630	Marine Engineering and Technology	The original code is 57050 and formerly known as "Marine Engineering".
4163010	Marine Engineering Structure and Construction	The original code is 5705010.
4163015	Seabed Mineral Development	The original code is 5705020.
4163020	Utilization of Seawater Resources	The original code is 5705030.
4163025	Marine Environmental Engineering	The original code is 5705040.
4163030	Coastal Engineering	
4163035	Offshore Engineering	
4163040	Deep Sea Engineering	
4163045	Marine Resources Development and Utilization Technology	Including Ocean Mineral Resources, Seawater Resources, Marine Biology, Marine Energy Development Technology, etc.
4163050	Ocean Observation and Forecasting Technology	Including Ocean Underwater Technology, Ocean Observation Technology, Ocean Remote Sensing Technology, Ocean Forecasting Technology, etc.
4163055	Marine Environmental Protection Technology	
4163099	Other Disciplines of Marine Engineering and Technology	The original code is 5705099.
420	Science and Technology of Surveying and Mapping	
42050	Marine Surveying and Mapping	
4205010	Marine Geodetic Survey	
4205015	Marine Gravity Survey	

Continued

Code	Disciplines	Descriptions
4205020	Marine Magnetic Survey	
4205025	Ocean Spring Layer Survey	
4205030	Ocean Sound Velocity Survey	
4205035	Hydrographic Survey	
4205040	Bathymetic Survey	
4205045	Hydrographic Charting	
4205050	Marine Engineering Survey	
4205099	Other Disciplines of Marine Surveying and Mapping	
480	Energy Science and Technology	
48060	Primary Energy	
4806020	Petroleum and Natural Gas Energy	
4806030	Hydroenergy	Including Ocean Energy, etc.
4806040	Wind Energy	
4806085	Natural Gas Hydrate Energy	
490	Nuclear Science and Technology	
49050	Nuclear Power Engineering Technology	
4905010	Marine Nuclear Power	
570	Hydraulic Engineering	
57010	Basic Disciplines of Hydraulic Engineering	
5701020	Rivers and Coastal Dynamics	
580	Transportation Engineering	
58040	Water Transport	
5804010	Navigation Technology and Equipment Engineering	Formerly known as "Nautical Navigation".
5804020	Ships Communication and Navigation Engineering	Formerly known as "Navigation Building & Navigation Mark Engineering".
5804030	Waterway Engineering	
5804040	Harbour Engineering	
5804080	Marine Technology and Equipment Engineering	
58050	Ships and Warships Engineering	
610	Environmental Science and Technology & Resource Science and Technology	
61020	Environmental Science	
6102020	Water Environment Science	Including Marine Environmental Science.
620	Safety Science and Technology	
62010	Basic Disciplines of Safety Science and Technology	
6201030	Catastrophology	Including Disaster Physics, Disaster Chemistry, Disaster Toxicology, etc.
780	Archaeology	
78060	Specialized Archaeology	

Continued

Code	Disciplines	Descriptions
7806070	Underwater Archaeology	
790	Economics	
79049	Resource Economics	
7904910	Marine Resource Economics	
830	Military Science	
83030	Science of Campaigns	
8303020	Science of Naval Campaigns	
83035	Science of Tactics	
8303530	Science of Naval Tactics	

Notes: According to percentage of the number of marine-related disciplines (third-level disciplines) that are included in the second-level disciplines in the total number of the third-level disciplines, the marine-related coefficient of proportionality of the second-level disciplines is determined as follows: acoustics (0.06), optics (0.06), geology (0.04), marine science (1), basic disciplines of fishery science (1), aquatic multiplication science (1), aquaculture science (1), aquatic feed science (1), aquatic protection science (1), fishing science (1), storage and processing of aquatic products (1), aquatic engineering (1), aquatic resource science (1), aquatic economics (1), other disciplines of fishery science (1), special medicine (0.33), systemic application of information technology (0.25), marine engineering and technology (1), marine surveying and mapping (1), primary energy (0.36), nuclear power engineering technology (0.20), basic disciplines of hydraulic engineering (0.25), water transport (0.56), ships and warships engineering (1), environmental science (0.17), basic disciplines of safety science and technology (0.17), specialized archaeology (0.11), resource economics (0.17), science of campaigns (0.17), science of tactics (0.17).

Compilation Explanations

In response to national marine innovation strategies and to offer service to the construction of national innovation system, the First Institute of Oceanography of State Oceanic Administration has been embarking on the measurement and calculation of marine innovation indicators from 2006 and officially initiated the research on national marine innovation index in 2013. *National Marine Innovation Index Report 2017-2018* is the fifth issue of related series of evaluation reports. Relevant information is explained as follows.

1. Needs Analysis

Innovation-driven development has become a national development strategy of our country. *Decision of the Central Committee of the Communist Party of China on Some Major Issues Concerning Comprehensively Deepening the Reform* explicitly put forward "building national innovation systems". Marine innovation is a key field in building an innovation-oriented country and an important component of national innovation system. Exploring and constructing the national marine innovation index and evaluating national marine innovation capability of our country are of great significance to the construction of a maritime power. The necessity to compile a series of evaluation reports of national marine innovation index is manifested in the following four aspects.

A. Urgent demand to comprehensively find out marine innovation situation in our country

To collect marine innovation data such as marine economic statistics, science and technology statistics and S&T achievements registration, so as to get a clear picture of the marine innovation situation of our country is the foundation of objective analysis of the marine innovation development trends of our country.

B. Objective demand to grasp the development trend of marine innovation in our country

To mine and analyze marine innovation data from the four aspects of marine innovation environment, marine innovation resources, marine knowledge creation, and marine innovation performance so as to gain a deeper knowledge of marine innovation and development trends in our country can meet the objective need of understanding the path and mode of marine innovation in China.

C. Actual demand to calculate accurately the key indicators of marine innovation in our country

To calculate and predict marine innovation key indicators such as the contribution rate of marine S&T progress, and the transformation rate of marine S&T achievements can give a realistic picture of the quality and efficiency of marine innovation in our country and provide a series of important indicators to support the formulation of marine innovation policy in China.

D. Practical demand to gain a comprehensive understanding of international marine innovation development

To analyze the development trend of international marine innovation and grasp the development situation of international marine innovation in basic research and technological R&D from the aspects of papers and patents in marine field so as to comprehensively understand the development trend of international marine innovation can provide a reference for the development of China's marine innovation.

2. Compilation Basis

A. The Report of the 19th National Congress of the Communist Party of China (CPC)

The Report of the 19th National Congress of the Communist Party of China clearly put forward the idea of "making China a country of innovators". It also pointed out that "innovation is the primary driving force behind development and the strategic support for the construction of a modern economic system". We should "aim for the frontiers of science and technology, strengthen basic research", "improve our national innovation system and boost our strategic scientific and technological strength" and "We will improve our national innovation system and boost our strategic scientific and technological strength".

B. The Report of the Fifth Plenary Session of the 18th Central Committee of the Communist Party of China

The Report of the Fifth Plenary Session of the 18th Central Committee of the Communist Party of China put forward that "we must give innovation top priority in overall national development, continue to promote innovations in theory, in institutions, in science and technology, in culture and in other fields, implement innovation throughout all the work of the Party and the state, and make innovation become common practice in the whole of society".

C. Outline of the National Strategy of Innovation-driven Development

The *Outline of the National Strategy of Innovation-driven Development* which was issued by the CPC Central Committee and the State Council in May, 2016, pointed out that "the implementation of the innovation-driven development strategy and the emphasis of S&T innovation as the strategic support for improving social productivity and comprehensive national strength put forward by the 18th National Congress of the CPC must be the focus of the overall national development. It is the important development strategy established by the CPC Central Committee at a new stage of development based on the overall situation, facing the world, focusing on the key points and promoting the development of the whole country".

D. Outline of the 13th Five-Year Plan for National Economic and Social Development of People's Republic of China

The *Outline of the 13th Five-Year Plan for National Economic and Social Development of People's Republic of China* put forward the major innovation-driven indicators and strengthened the leading role of S&T innovation. It also pointed out that "development must be based on innovation with S&T innovation as focus and talents development as support to promote the organic combination of S&T innovation with mass entrepreneurship and innovation, and create more leading development models driven by innovation and give more play of first mover advantage".

E. Vision and Actions on Jointly Building a Silk Road Economic Belt and 21st-Century Maritime Silk Road

The *Vision and Actions on Jointly Building a Silk Road Economic Belt and 21st-Century Maritime Silk Road* put forward new development ideas of "innovating and opening up the economic systems and mechanism, intensifying scientific and technological innovation, developing a new advantage in participating and leading the international cooperation and competition, and becoming the pace-setter and main force in the construction of the Belt and Road, particularly the building of the 21st-Century Maritime Silk Road.

F. Decision of the Central Committee of the Communist Party of China on Major Issues Concerning Comprehensively Deepening the Reform

Decision of the Central Committee of the Communist Party of China on Major Issues Concerning Comprehensively Deepening the Reform explicitly stated "the building of a national innovation system".

G. The 13th Five-Year Plan on Science, Technology and Innovation

the 13th Five-Year Plan on science, Technology and Innovation put forward that "the period of the 13th Five-Year Plan is the decisive stage for building a moderately prosperous society in all aspects and stepping into innovative countries. It is also the key stage of implementing innovation-driven development strategies deeply and deepening the reform of scientific and technological system across the board. Facing the world and based on the whole situation, we must conscientiously carry out the decisions and deployment of the CPC Central Committee and the State Council, thoroughly understand and accurately grasp the new requirements in the context of new normal of economic growth and new trend of S&T innovation both at home and abroad. We need to explore the new path of innovation development systematically, open up new realm of development led by S&T innovation, stride into the ranks of innovative countries, and accelerate the construction of world-class S&T power".

H. Overall Planning of Marine Scientific and Technological Innovation

At the first working conference on strategic research, the *Overall Planning of Marine Scientific and Technological Innovation* urged that efforts be made "to conduct marine strategy research on the basis of 'overall planning' and 'innovation'" and "to recognize the innovation path and modes and to take stock of 'overall resources'".

I. The 13th Five-Year Special Plan for Scientific and Technological Innovation in the Marine Field

The 13th Five-Year Special Plan for Scientific and Technological Innovation in the Marine Field stated clearly that "we should further build and improve the national marine S&T innovation system, enhance China's marine S&T innovation capability, and significantly enhance the supporting role of scientific and technological innovation in improving the development of marine industry".

J. Outline of the National Marine Economy Development Plan

The *Outline of the National Marine Economy Development Plan* proposed that we should "gradually build China into a maritime power".

K. Outline of Marine Development by Means of Science and Technology (2016-2020)

The *Outline of Marine Development by Means of Science and Technology (2016-2020)* put forward that "a long-term mechanism for the development of the ocean by means of science and technology that is conducive to innovation-driven development will have been formed and a system of transfer and transformation of marine S&T achievements with chain-type layout, complementary advantages, collaborative innovation, and agglomeration and transformation will have been constructed by 2020. The capability of marine science and technology to lead the sustainable growth of marine biomedicine and products, marine high-end equipment manufacturing, seawater desalination and comprehensive utilization will have been notably enhanced. The capability of cultivating new industries such as marine new materials, marine environmental protection, and modern marine services will have been continuously strengthened. The ability to support integrated ocean management and public service will have improved significantly. The transformation rate of marine S&T achievements will have exceeded 55%, the contribution rate of marine S&T progress to marine economic growth will have exceeded 60%, the average annual growth rate of invention patents owner ship will have reached 20%, and the self-sufficiency rate of marine high-end equipment will have reached 50%. A situation in which the marine economy and the marine industry interact and develop in an integrated manner will have been formed, laying a solid foundation for the construction of a maritime power and China's entry into an innovative country".

L. National Medium and Long-Term Program for Science and Technology Development (2006-2020)

The *National Medium and Long-Term Program for Science and Technology Development (2006-2020)* put forward that "we must place the improvement of independent innovative capability at the core in adjusting economic structure, transforming growth mode, and improving national competitiveness. Building an innovation-oriented country is therefore a major strategic choice for China's future development". It also pointed out the guiding principles for our S&T undertakings which are "independent innovation, leapfrogging in key fields, supporting development, and leading the future", and emphasized that "we should advance the construction of national innovation system with Chinese characteristics in all-round way and drastically enhance the nation's independent innovation capability".

3. Data Sources

The data used by *National Marine Innovation Index Report 2017-2018* are mainly from the following sources.

a. *China Statistical Yearbook.*

b. *China Marine Statistical Yearbook.*

c. Statistical data of the Ministry of Science and Technology.

d. Statistical data of marine-related higher institutions and marine-related disciplines of the Ministry of Education.

e. Data on marine science papers and marine patents from Lanzhou Information Center, Chinese Academy of Sciences.

f. Database of Chinese Science Citation Database, CSCD.

g. Database of Science Citation Index Expanded, SCIE.

h. Derwent Innovations Index, DII.

i. Engineering Index Compendex, EI.

j. Registration data on marine S&T achievements.

k. *Compendium of Science and Technology Statistics in Institutions of Higher Education.*

l. Other publications.

4. Compilation Process

Marine Policy Research Center, First Institute of Oceanography, MNR organized the compilation of *National Marine Innovation Index Report 2017-2018*. Lanzhou Information Center, Chinese Academy of Sciences participated in the compilation of marine papers, patents and thematic analysis of international ocean science and technology research. Department of Inovation and Development, Ministry of Science and Technology of the People 'sRepublic of China, Science and Technology Department of the Ministry of Education, National Marine Data and Information Service, School of Management of Huazhong University of Science and Technology and other units and departments provided data support. The specific compilation process is divided into three stages, which are the preparation stage, data calculation and report preparation stage, and finally the consultation and revision stage. Specific introductions are as follows.

A. Preparation Stage

The formation of basic ideas from January to February, 2018. The previous four series of reports on national marine innovation, namely the *Tentative Evaluation Report of National Marine Innovation Index 2013*, *Tentative Evaluation Report of National Marine Innovation Index 2014*, *National Marine Innovation Index Report 2015*, and *National Marine Innovation Index Report 2016* were published in May 2015, December 2015, December 2016 and January 2018 respectively. On the basis of the previous work and after many rounds of research and communications to summarize experiences and find out deficiencies of the previous four issues, in early 2018, we formed the ideas on compiling *National Marine Innovation Index Report 2017-2018* and made a specific compiling plan.

The collection of data. In January 2018, we smoothly obtained S&T innovation data of marine scientific research institutes, related data of *Compendium of Science and Technology Statistics in Institutions of Higher Education* and marine-related S&T innovation data extracted by marine-related higher education institutions based on marine-related disciplines (Level-one) from Science and Technology Statistics Information Center of Huazhong University of Science and Technology and Science and Technology Department of the Ministry of Education. Meanwhile, working with Lanzhou Information Center, Chinese Academy of Sciences, we collected data for SCI papers in the marine field and marine patents, and so on.

Building of the report compiling team and the indicator calculation team. In January 2018, on the basis of the original compiling teams of *National Marine Innovation Index Report 2016*, we built the compiling and indicator calculation teams of *National Marine Innovation Index Report 2017-2018*, which are composed of staff from Marine Policy Research Center, the First Institute of Oceanography, MNR and Lanzhou Information Center, Chinese Academy of Sciences.

B. Data Calculation and Report Preparation Stage

Data processing and analysis. From January to February 2018, we have processed and analyzed the S&T innovation data of marine scientific research institutes and data from *China Statistical Yearbook, China Marine Statistical Yearbook, Compendium of Science and Technology Statistics in Institutions of Higher Education*, and marine-related S&T innovation data extracted by marine-related higher education institutions based on marine-related disciplines (Level-one).

Data calculation. From February 20 to March 20, 2018, the calculation team has calculated the contribution rate of marine S&T progress and transformation rate of marine S&T achievements. National marine innovation index and regional marine innovation index have also been calculated according to the corresponding evaluation methods.

Compilation of the first draft of the report. From March 21 to April 20, 2018, the first draft of the report was completed based on the results of data analysis and indicator calculation.

The first round of data review. From April 21 to May 7, 2018, the calculation team has conducted the first round of data review, focusing on data sources, data processing procedures, figures and tables.

Revision of the second draft of the report. From May 8 to 22, 2018, the first draft of the report has been revised according to the results of data review and the calculation results of indicators, and the second draft was completed.

The second round of data recheck. From May 23 to 31, 2018, the calculation team has taken the reverse review method and conducted the second round of data review, reviewing figures, tables, data processing, and data sources one by one according to the report content.

Solicitation of opinions on a small scale. From June 1 to 8, 2018, internal opinions have been solicited on a small scale.

The third round of data recheck. From June 1 to 5, 2018, data sources, data processing process, and the correspondence between texts and charts have been rechecked by combining a forward review method with reverse review method.

Improvement of the third draft of the report. From June 9 to 14, 2018, the second draft of the report has been improved according to the results of the third round of data review and opinions solicited on a small scale, and the third draft was completed.

C. Stages of Report Review, Revision and Improvement

Revision according to the feedback of experts. On June 15, 2018, an expert consultation meeting was held to solicit opinions from experts. The report was revised according the suggestions from the experts.

Internal review and revision of the fourth draft of the report. From July 26 to August 2, 2018, the center has organized internal review and revised the report according to the suggestions of the internal review.

Review of administration departments. From July 26 to August 4, 2018, the report was submitted to the Department of Science and Technology Development of the Ministry of Natural Resources[1] of the People's Republic of China and the Department of Innovation and Development of the Ministry of Science

---

[1] Original: Department of Science and Technology of State Oceanic Administration

and Technology of the People's Republic of China for review and revised according to the suggestions of that review.

Recheck of calculation process. From July 30 to August 4, 2018, the calculation team has rechecked the calculation process, focusing on formulas of calculation, parameters and accuracy of results. The report has been improved and perfected according to the results of the recheck and the fourth draft was completed.

Review of advisory group. In August 2018, the advisory group reviewed the report and the report was revised according to the suggestions of the advisory group.

Text proofreading. From September 1 to 25, 2018, the compiling team has proofread the report according to the chapter and revised the text according to the opinions and suggestions of each member.

Pre-editing of Science Press. In September 2018, the electronic version of the report was submitted to the editorial department of Science Press for pre-editing.

Examination of administration departments. On October 19, 2018, it was submitted to the Department of Science and Technology Development of the Ministry of Natural Resources of the People's Republic of China for examination.

5. Adoption of Opinions and Suggestions

More than 30 persons have been solicited for opinions for more than 30 times and over 300 pieces of opinions and suggestions have been received after summary.

220 pieces of opinions and suggestions have been adopted, with the adoption rate of feedback and suggestions being approximately 73.3%.

Instructions on Updates

1. Some Chapters and Contents Have Been Added and Deleted

A. Chapter V "Spatial Distribution Characteristics and Evolutionary Trend of China's Marine Scientific Research Institutions" and Chapter VI "Analysis of Global Marine Innovation Capability" are newly added.

B. Chapter VI "Specific Analysis on Regional Demonstration of China's Marine Economic Innovation Development" and Chapter VII "Specific Analysis of the Input-output Efficiency of China's Marine Science and Technology" of *National Marine Innovation Report 2016* are deleted.

2. Both Domestic and International Data Have Been Updated

a. The data of international marine-related innovation papers are updated. The original data are updated as of 2016, used to analyze marine innovation output achievements and make comparative analysis of domestic and foreign marine innovation papers.

b. The data of international patents are updated. The original data are updated as of 2016, used to analyze marine innovation output achievements and make comparative analysis of domestic and foreign marine innovation patents.

c. The domestic data are updated. The original data used by the evaluation indicators of national marine innovation are updated as of 2016, and the evaluation indicators of the regional marine innovation sub-index are updated as of 2016.

d. The sources of data are updated. Since the quantity of enterprise data provided is lower than the quantity collected in previous years, the calculation and comparison of the corresponding data are greatly affected. Therefore, the enterprise data are excluded and new raw data are formed. The re-calculated index will have a corresponding gap with the data in the previous report.

3. The Indicator System of National Marine Innovation Index Has Been Adjusted

Because the data of enterprises are no longer included in the science and technology statistics of the Ministry of Science and Technology in 2016, the sub-index of the indicator system of national marine innovation index is adjusted, and the sub-index of marine enterprise innovation is deleted.

第一章　从数据看我国海洋创新

在海洋强国和"一带一路"倡议背景下，我国海洋创新发展不断取得新成就，部分领域达到国际先进水平，海洋创新条件和环境条件明显改善。

海洋创新人力资源持续优化。海洋科研机构中科技活动人员结构持续改善，R&D（科学研究与试验发展，research and development）人员总量、折合全时工作量稳步上升，R&D人员学历结构不断优化。

海洋创新国家级平台保持稳定。海洋科研机构的国家（重点/工程）实验室和国家工程（研究/技术研究）中心数量近年基本稳定，海洋科研机构的基本建设与固定资产逐年增加。

海洋创新经费规模显著提升。海洋科研机构的R&D经费规模显著提升，R&D经费内部支出稳定增长。

海洋创新产出成果稳步增长。海洋科研机构的海洋科技论文总量保持增长，海洋领域SCI论文发表数量大幅增长，海洋科技著作出版种类明显增长，专利申请量、授权量涨势强劲。

高等学校海洋创新能力稳步提升。涉海高等学校的人员、经费、课题等方面均呈现逐年增长的态势。

海洋科技对海洋经济发展贡献稳步增强。2016年海洋科技进步贡献率达到65.9%[1]，海洋科技成果转化率达到50.0%[2]，海洋科技创新促进成果转化的作用日益彰显。

---

[1] 2016年海洋科技进步贡献率根据2006～2016年相关数据测算所得
[2] 2016年海洋科技成果转化率根据2000～2016年相关数据测算所得

第一节　海洋创新人力资源结构稳定

海洋创新人力资源是建设海洋强国和创新型国家的主导力量与战略资源，海洋创新科研人员的综合素质决定了国家海洋创新能力提升的速度和幅度。海洋科研机构的科技活动人员和 R&D 人员是重要的海洋创新人力资源，突出反映了一个国家海洋创新人才资源的储备状况。其中，科技活动人员是指海洋科研机构中从事科技活动的人员，包括科技管理人员、课题活动人员和科技服务人员；R&D 人员是指海洋科研机构本单位人员和外聘研究人员，以及在读研究生中参加 R&D 课题的人员、R&D 课题管理人员和为 R&D 活动提供直接服务的人员。

一、科技活动人员结构持续优化

从人员组成上来看，2011～2016 年我国海洋科研机构课题活动人员(即编制在研究室或课题组的人员)在科技活动人员中的占比保持在 64%以上，2016 年有较大幅度提升；科技管理人员(即机构领导及业务、人事管理人员)和科技服务人员(即直接为科技工作服务的各类人员)所占比例大部分在 15%以下，但 2014～2016 年科技服务人员所占比例相对较高(图 1-1)。从人员学历结构上来看，近 6 年来，我国海洋科研机构科技活动人员中博士、硕士毕业生占比总体呈增长态势，2016 年博士、硕士毕业生分别占科技活动人员总量的 28.14%和 32.32%，均比 2015 年有所提升(图 1-2)。从人员职称结构上来看，2011～2016 年，我国海洋科研机构科技活动人员中高级、中级职称人员占比保持在初级职称人员占比的 2 倍左右，2016 年高级、中级职称人员分别占科技活动人员总量的 41.85%和 34.15%(图 1-3)。

二、R&D 人员总量、折合全时工作量稳中有升

2002～2016 年，我国海洋科研机构的 R&D 人员总量和折合全时工作量总体呈现稳步上升态势(图 1-4)。2002～2006 年，R&D 人员总量和折合全时工作量增长相对较缓；2006～2007 年，两者均涨势迅猛，增长率分别为 119.1%和 88.16%；2007～2014 年，两者保持稳步增长；2014～2015 年，R&D 人员总量略有下降；2015～2016 年，两者再次出现明显增长，增长率分别为 13.68%和 6.55%。

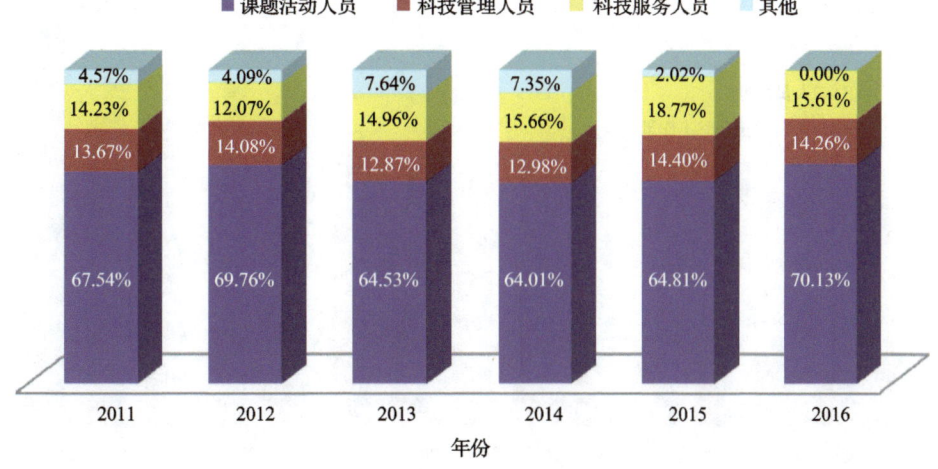

图 1-1　2011～2016 年海洋科研机构科技活动人员构成①

---

① 图中数据百分比之和不等于 100%是因为有些数据进行过舍入修约

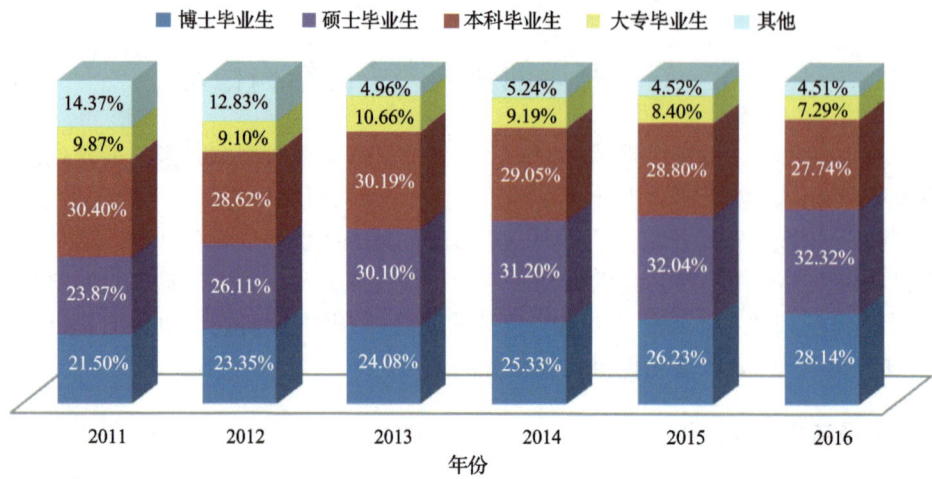

图 1-2　2011～2016 年海洋科研机构科技活动人员学历结构

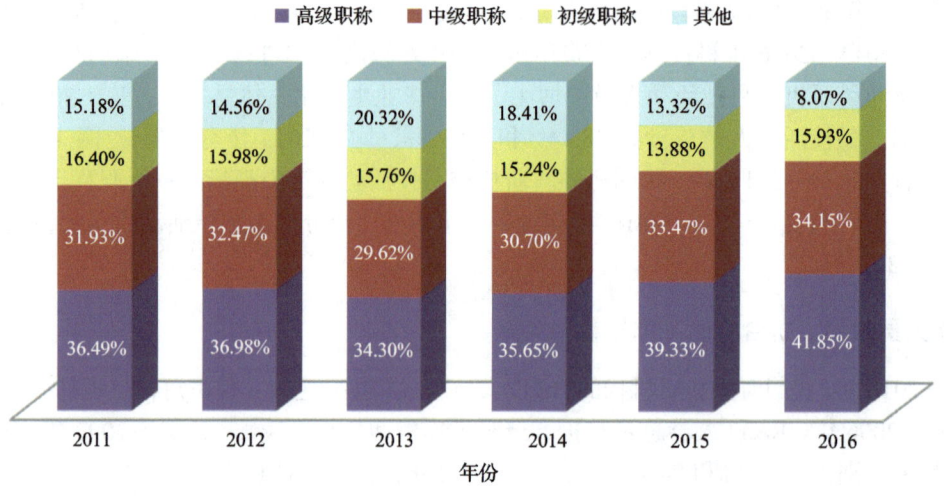

图 1-3　2011～2016 年海洋科研机构科技活动人员职称结构

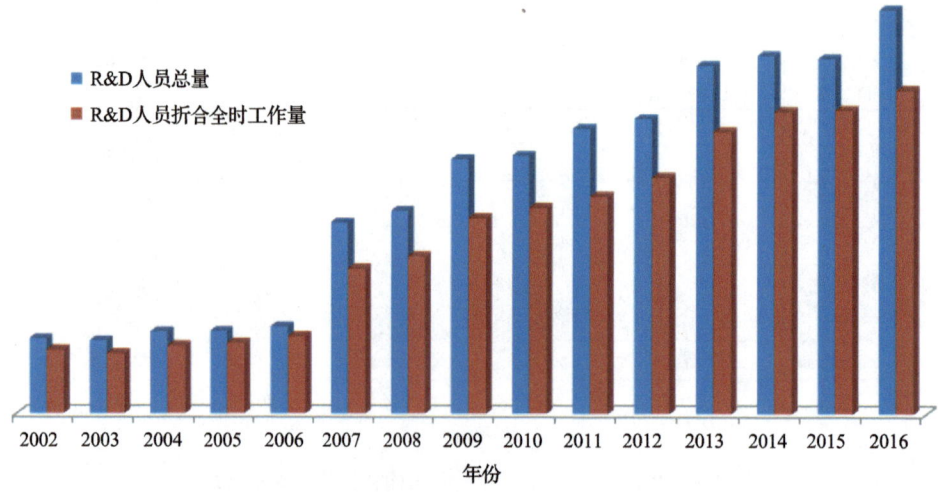

图 1-4　2002～2016 年海洋科研机构 R&D 人员总量（人）、折合全时工作量（人·年）变化趋势

三、R&D 人员学历结构基本稳定

2011～2016 年，我国海洋科研机构 R&D 人员中博士毕业生数量保持增长，占比呈波动上升趋势，硕士毕业生数量整体呈现增长态势。2016 年博士和硕士毕业生分别占 R&D 人员总量的 31.67% 和 32.97%（图 1-5）。其中，博士毕业生占比 2015 年最高，达到 31.99%，相比 2011 年增长 4.17%；硕士毕业生占比呈波动增长态势，2016 年相比 2011 年增长 6.07%。

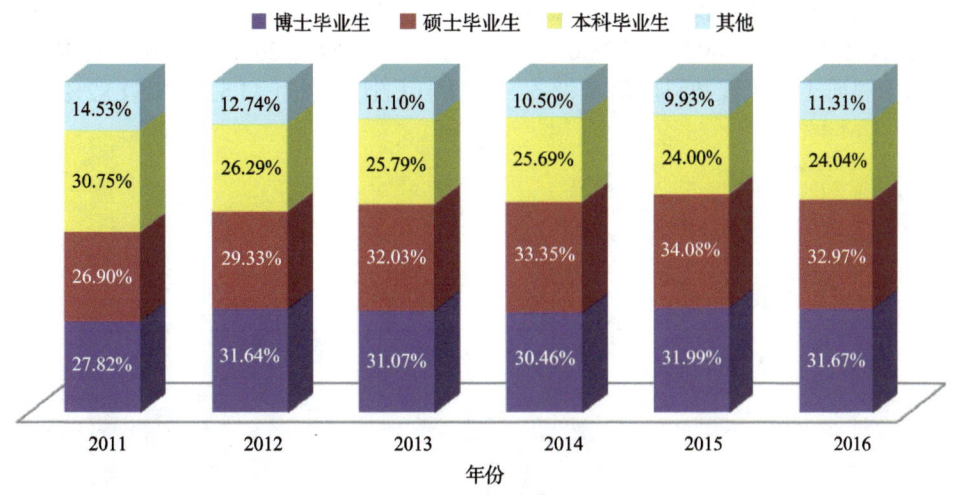

图 1-5　2011～2016 年海洋科研机构 R&D 人员学历结构

第二节　海洋创新平台环境逐渐改善

一、海洋科研机构的国家(重点/工程)实验室和国家工程(研究/技术研究)中心数量总体增长趋势

2002～2016 年，国家(重点/工程)实验室和国家工程(研究/技术研究)中心数量总体增长趋势，2016 年比 2015 年有所下降，国家(重点/工程)实验室个数 2010 年达到最大值，国家工程(研究/技术研究)中心个数在 2013～2015 年维持最大值（图 1-6）。

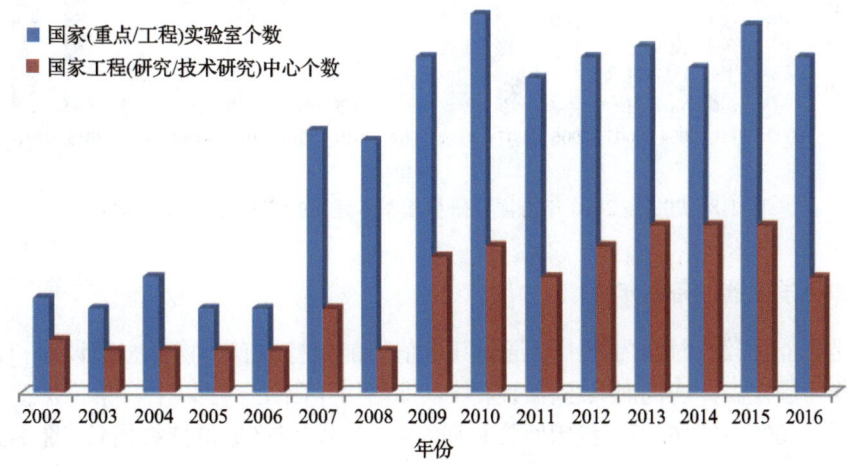

图 1-6　2002～2016 年海洋科研机构国家(重点/工程)实验室和国家工程(研究/技术研究)中心数量(个)趋势

二、基本建设投资实际完成额保持增长

基本建设投资实际完成额是指本机构在当年完成的用货币表示的基本建设工作量,按用途分为科研仪器设备、科研土建工程、生产经营土建与设备和生活土建与设备。基本建设投资科研仪器设备是指在基本建设投资的实际完成额中购置的科研仪器设备总值,基本建设投资科研土建工程是指在基本建设投资的实际完成额中完成的科研土建工作量(如科研楼、试验用房等)。2002~2016年,我国海洋科研机构的基本建设投资实际完成额呈增长态势(图1-7),2007年增长最为迅猛,年增长率达到228.66%,2016年是2002年的28.66倍。从用途分类来看,2002~2016年基本建设投资实际完成额主要用于科研土建工程和科研仪器设备,2016年占比分别为44.52%和54.82%(图1-8)。

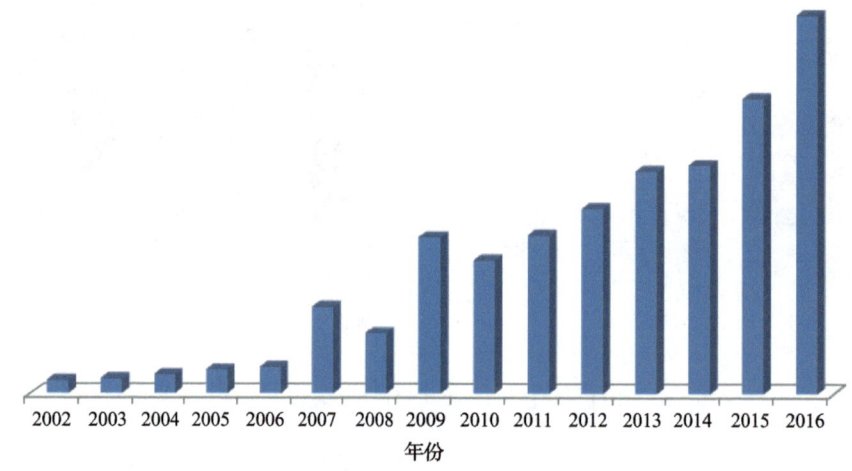

图1-7 2002~2016年海洋科研机构基本建设投资实际完成额(千元)变化趋势

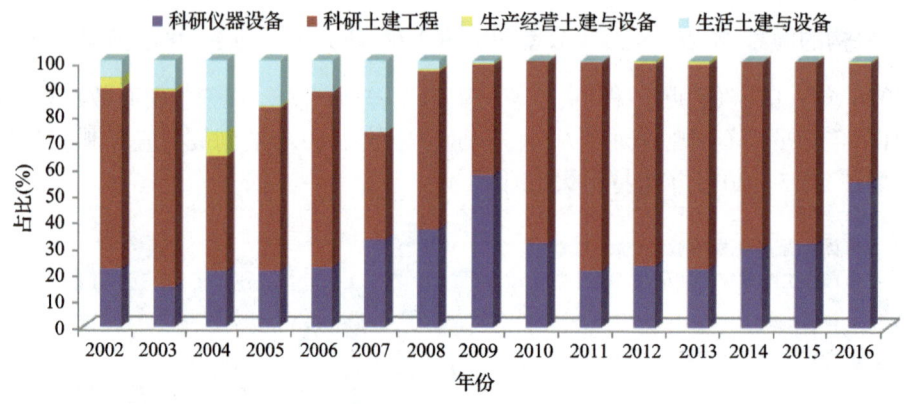

图1-8 2002~2016年海洋科研机构基本建设投资实际完成额构成

三、固定资产和科学仪器设备逐年递增

固定资产是指能在较长时间内使用,消耗其价值但能保持原有实物形态的设施和设备,如房屋和建筑物等。作为固定资产应同时具备两个条件,即耐用年限在一年以上,单位价值在规定标准以上的财产、物资。2002~2016年,我国海洋科研机构的固定资产原价持续增长(图1-9),年均增长率为22.04%。固定资产原价中科学仪器设备是指从事科技活动的人员直接使用的科研仪器设备,不包括与基建配套的各种动力设备、机械设备、辅助设备,也不包括一般运输工具(科学考察用交通运输工具除外)和专用于生产的仪器设备。2002~2016年,我国海洋科研机构固定资产原价中科学仪

器设备部分同样保持增长态势(图 1-9),年均增长率为 24.88%。

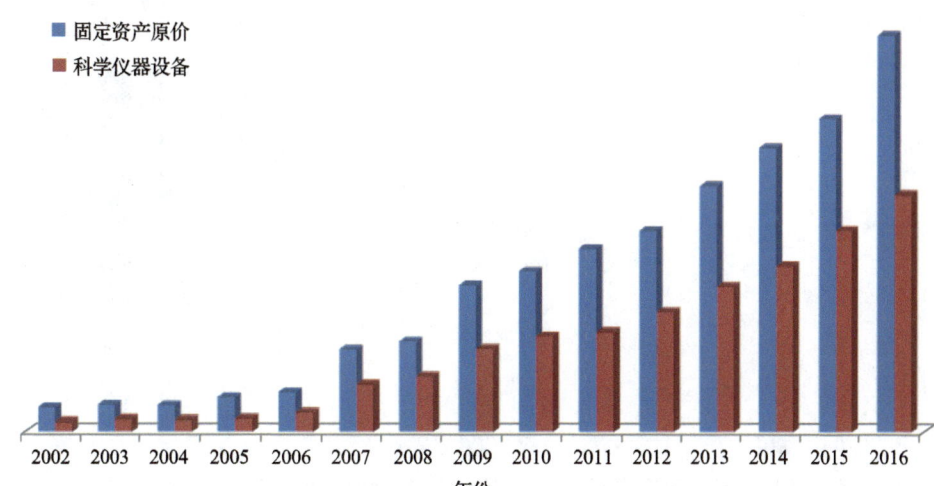

图 1-9　2002~2016 年海洋科研机构固定资产原价(千元)和固定资产原价中科学仪器设备(千元)变化趋势

第三节　海洋创新经费规模显著提升

R&D 活动是创新活动最为核心的部分,不仅是知识创造和自主创新能力的源泉,还是全球化环境下吸收新知识和新技术的能力基础,更是反映科技经济协调发展和衡量经济增长质量的重要指标。海洋科研机构的 R&D 经费是重要的海洋创新经费,能够有效反映国家海洋创新活动规模,客观评价国家海洋科技实力和创新能力。

一、R&D 经费规模稳中有升

2002~2016 年,我国海洋科研机构的 R&D 经费总体保持增长态势(图 1-10),年均增长率达到 23.72%。2007 年是该指标迅猛增长的一年,年增长率达到 145.18%。

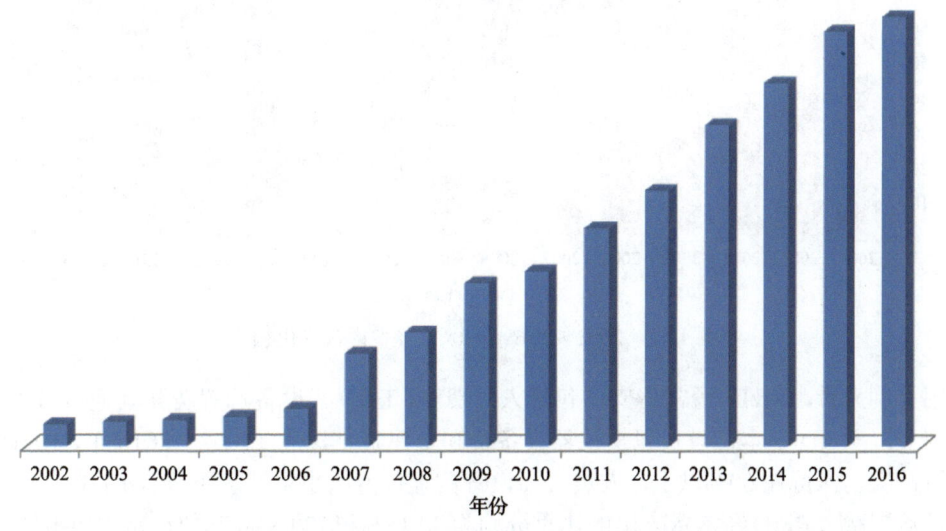

图 1-10　2002~2016 年 R&D 经费(千元)变化趋势

R&D 经费占全国海洋生产总值的比重通常作为国家海洋科研经费投入强度指标,反映国家海洋创新资金投入强度。2002~2016年,该指标整体呈现增长态势,年均增长率为8.54%;与2015年相比,2016年该指标略有下降(图1-11)。

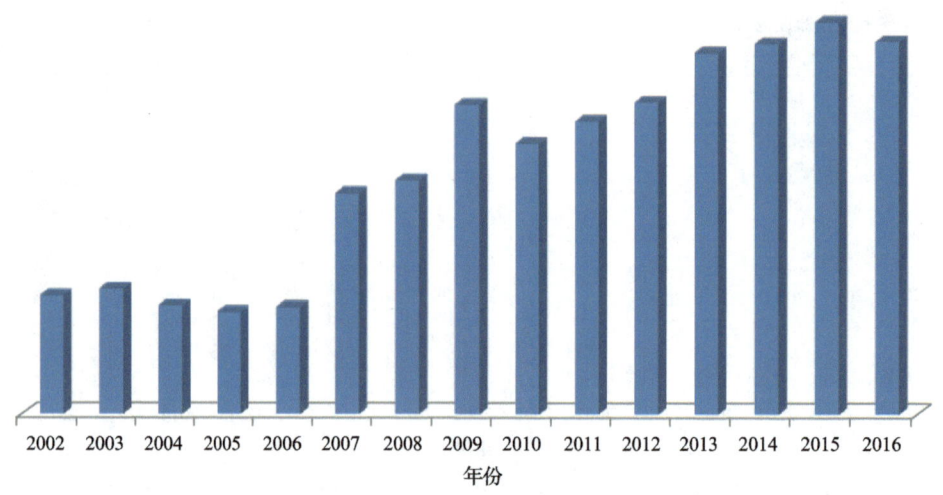

图1-11 2002~2016年R&D经费占全国海洋生产总值的比重变化趋势

二、R&D经费内部支出稳定增长

R&D经费内部支出是指当年为进行R&D活动而实际用于机构内的全部支出,包括R&D经常费支出和R&D基本建设费。2002~2016年,R&D基本建设费在R&D经费内部支出中的占比波动上升(图1-12),占比从2002年的8.71%上升到2016年的13.98%,体现了我国对基建投资重视程度的提高。

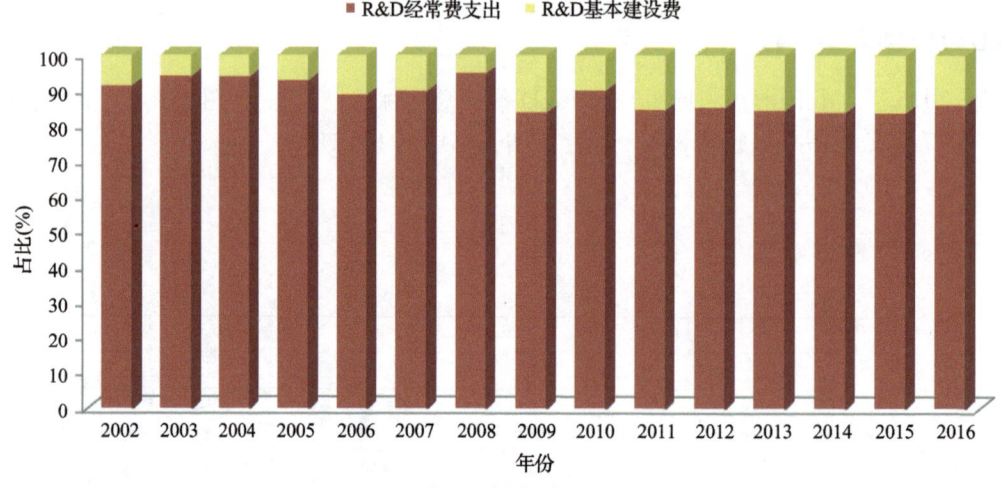

图1-12 2002~2016年R&D经费内部支出构成

从费用类别来看,R&D经常费支出包括人员费(含工资)、设备购置费和其他日常支出(包括业务费和管理费),R&D基本建设费包括仪器设备费和土建费。其中,2002~2016年,R&D经常费支出中其他日常支出保持在50%以上,人员费和设备购置费占比小幅波动下降(图1-13),2016年,人员费和设备购置费占R&D经常费支出的比重分别为30.15%和12.08%,其他日常支出占比为57.77%。2002~2016年,R&D基本建设费构成呈波动趋势,除2007年和2009年土建费占比小于仪器设备费占比外,其他年份均超过仪器设备费占比(图1-14)。

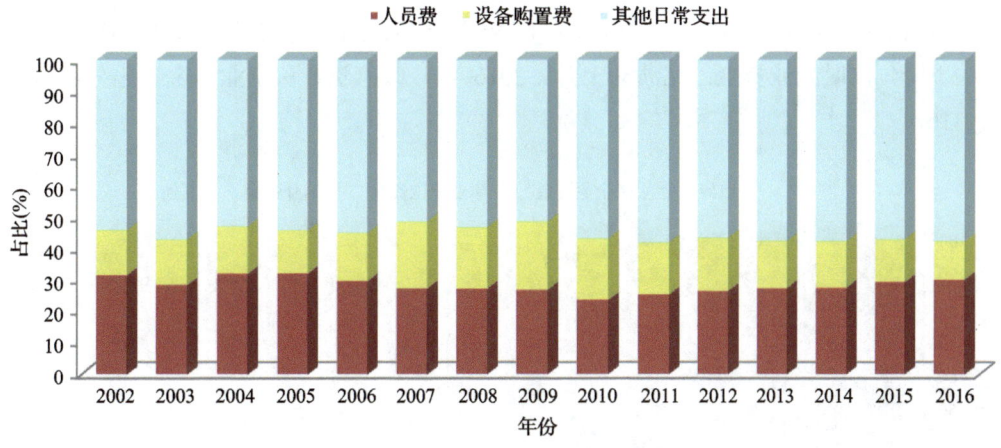

图 1-13 2002~2016 年 R&D 经常费支出构成(按费用类别)

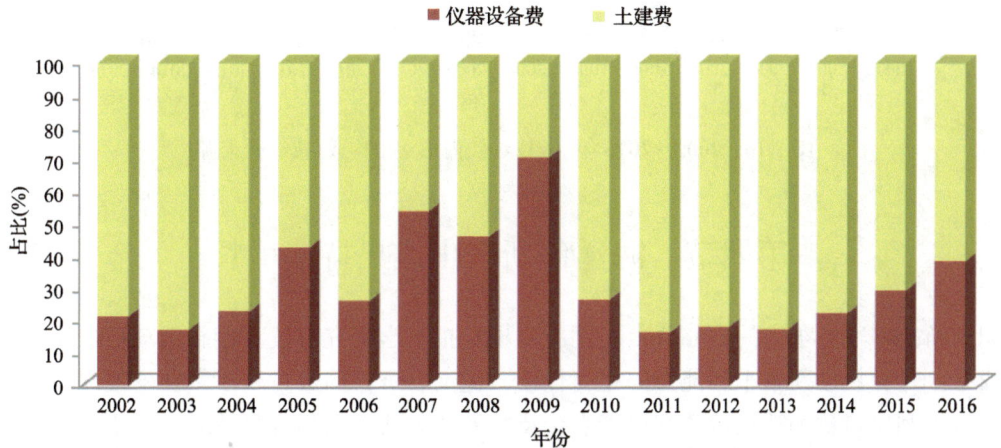

图 1-14 2002~2016 年 R&D 基本建设费构成(按费用类别)

从活动类型来看,2002~2016 年,R&D 经常费支出中用于基础研究的经费占比总体呈波动上升趋势,用于应用研究的经费占比从 2002 年的 48.73%下降至 2016 年的 37.73%,用于试验发展的经费占比从 2002 年的 33.05%下降至 2016 年的 31.28%(图 1-15)。

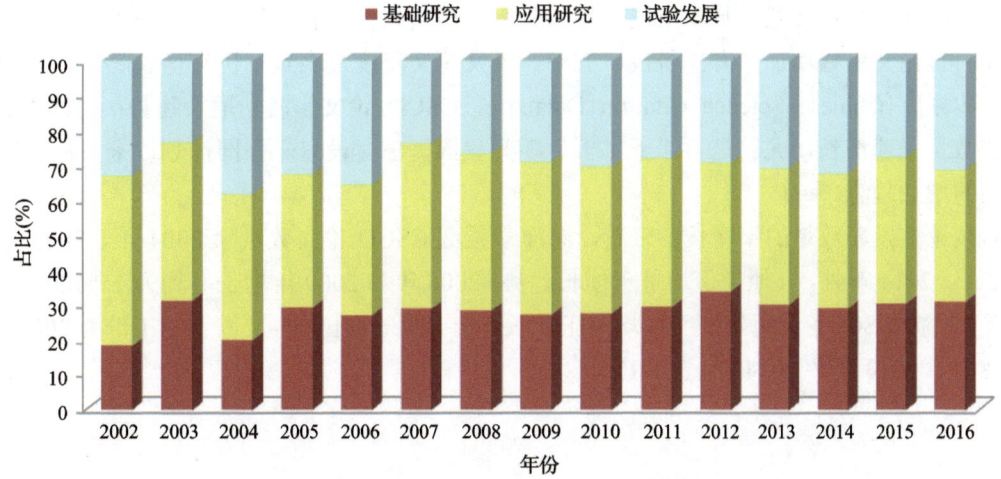

图 1-15 2002~2016 年 R&D 经常费支出构成(按活动类型)

从经费来源来看，2002~2016 年，R&D 经费内部支出主要来源于政府资金和企业资金，且政府资金占比波动下降，同时企业资金占比波动上升。2016 年，政府资金和企业资金占比分别为 80.69%和 9.11%（图 1-16）。

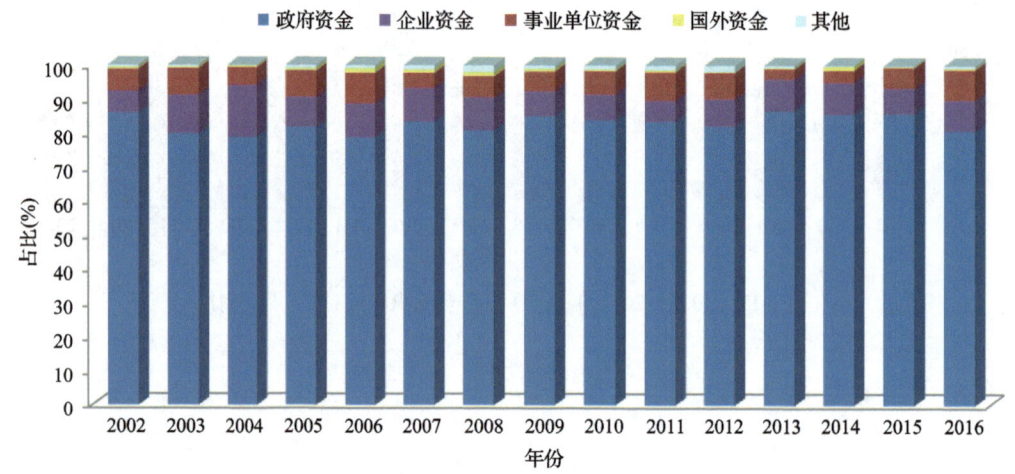

图 1-16　2002~2016 年 R&D 经费内部支出构成（按经费来源）

第四节　海洋创新产出成果持续增长

知识创新是国家竞争力的核心要素。创新产出是指科学研究与技术创新活动所产生的各种形式的中间成果，是科技创新水平和能力的重要体现。论文、著作的数量和质量能够反映海洋科技原始创新能力，专利申请量和授权量等则更能直接反映海洋创新活动程度和技术创新水平。较高的海洋知识扩散与应用能力是创新型海洋强国的共同特征之一。

一、海洋科技论文总量保持增长

海洋科技论文总量保持稳定增长态势。2001~2016 年，我国海洋学领域科技论文总量持续增长，2016 年论文发表数量是 2001 年的 6.06 倍，年均增长率为 12.77%。如图 1-17 所示，"十一五"到"十二五"期间海洋科技论文数量基本呈线性增长趋势，但增长幅度有差异，尤其是"十二五"期间，我国海洋科技论文迅猛增加。海洋科技研究中、外文论文年度发表数量也有增长趋势，其中，中国科学引文数据库（Chinese Science Citation Database，CSCD）论文呈波动增长趋势；海洋学领域 SCI 论文发表数量飞速增长，尤其是自"十二五"期间我国提出建设海洋强国计划以来，论文发表数量呈现明显的增长趋势。

从科技论文发表数量的年增长率来看，海洋学领域 CSCD 论文数量除 2004 年、2005 年、2008 年、2012 年、2014 年外，其他年份均呈正增长趋势，2006 年和 2009 年增长率均为 15%以上；除 2005 年外，海洋学领域 SCI 论文每年发文量均为正增长趋势，增长率在 25%及以上的年份为 2003 年、2004 年、2008 年、2013 年和 2014 年（表 1-1）。

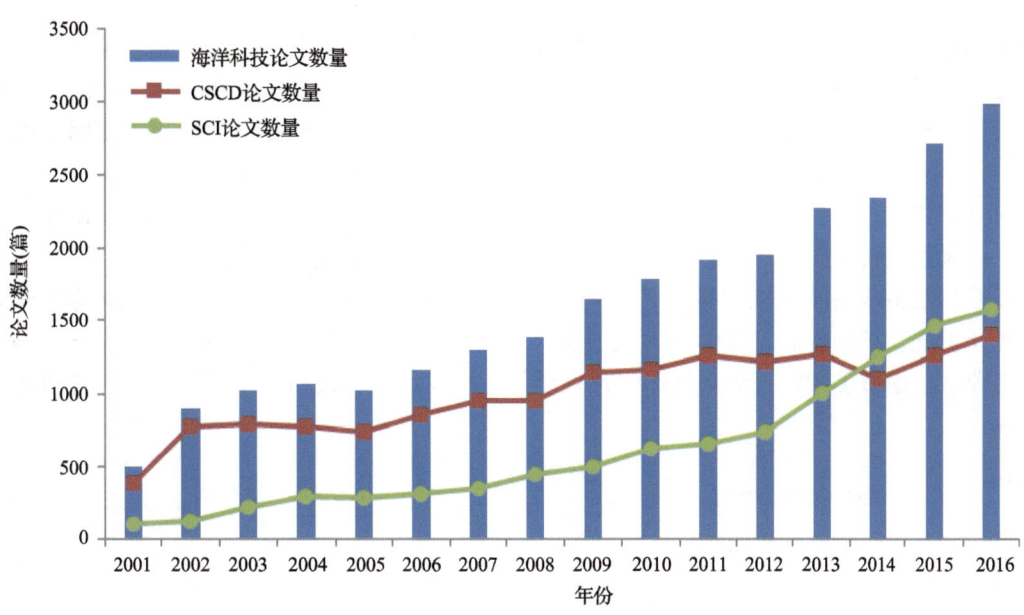

图 1-17　2001～2016 年我国海洋科技论文发表数量年度变化

表 1-1　2001～2016 年我国海洋科技论文发表数量及年增长率分析

年份	CSCD 论文数量（篇）	SCI 论文数量（篇）	海洋科技论文数量（篇）	年增长率(%)	
				CSCD 论文数量	SCI 论文数量
2001	384	108	492		
2002	767	126	893	100	17
2003	791	224	1015	3	78
2004	772	294	1066	−2	31
2005	737	279	1016	−5	-5
2006	853	308	1161	16	10
2007	948	346	1294	11	12
2008	943	442	1385	−1	28
2009	1144	499	1643	21	13
2010	1161	619	1780	1	24
2011	1257	655	1912	8	6
2012	1217	736	1953	−3	12
2013	1265	1000	2265	4	36
2014	1097	1246	2343	−13	25
2015	1254	1461	2715	14	17
2016	1403	1580	2983	12	8

二、我国海洋学 SCI 论文量质齐升

2001～2016 年我国发表海洋学 SCI 论文数量为 9923 篇，呈现明显增长趋势，尤其是 2012 年之后快速增长（图 1-18），2016 年发表论文数量是 2001 年的 14.63 倍。2013 年是 SCI 论文增长量的突变年。SCI 论文增长量在 2006～2010 年呈现波动增长趋势，2011～2016 年 SCI 先增后减。如

图 1-19 所示，2001~2016 年，国际发表海洋学 SCI 论文数量有增有减，我国发表论文数量呈现持续增长趋势，尤其是进入"十二五"之后，呈现快速增长趋势。如图 1-20 所示，2001~2016 年，我国作为第一作者国家署名的 SCI 论文数量呈现增长趋势，其中 2012~2015 年呈现明显的直线增长趋势。

我国海洋领域 SCI 论文学科交叉频繁，共覆盖 252 个学科类别。根据检索式，在 Web of Science 数据库中检索到的我国海洋科技相关的 SCIE 研究论文共涉及 22 种学科类别，如表 1-2 所示，海洋科技研究涉及众多学科领域且学科之间交叉频繁。除海洋学外，在研究成果中涉及较多的学科领域还包括海洋工程、湖沼学、气象学与大气科学、水资源学、海洋与淡水生物学、生态学、地球化学与地球物理学、地质工程、采矿与选矿、渔业学等，以及海洋科技相关学科和交叉学科领域。

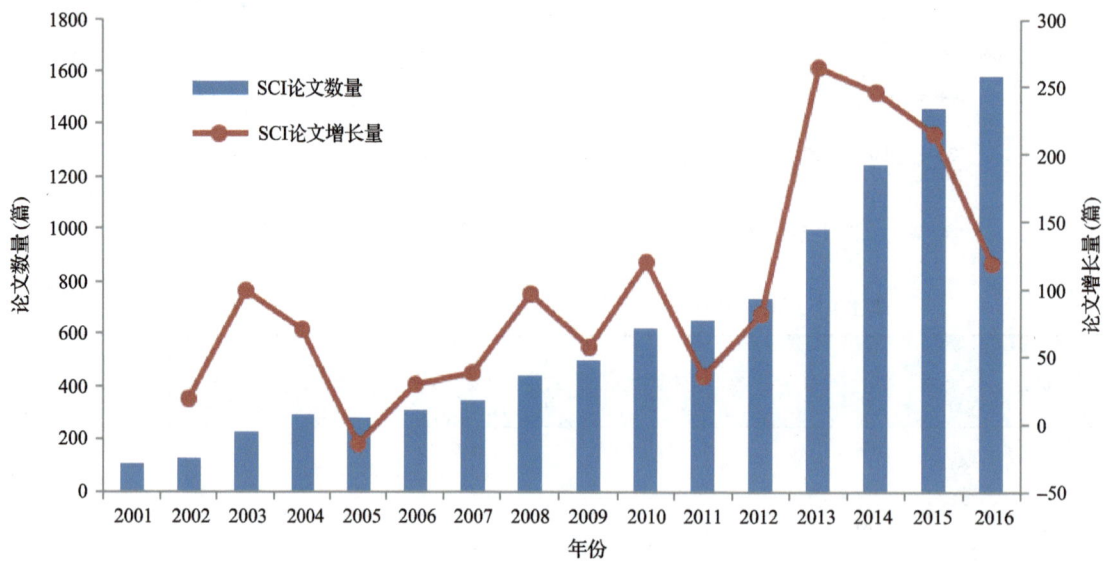

图 1-18　2001~2016 年我国发表海洋学 SCI 论文数量及增长量变化

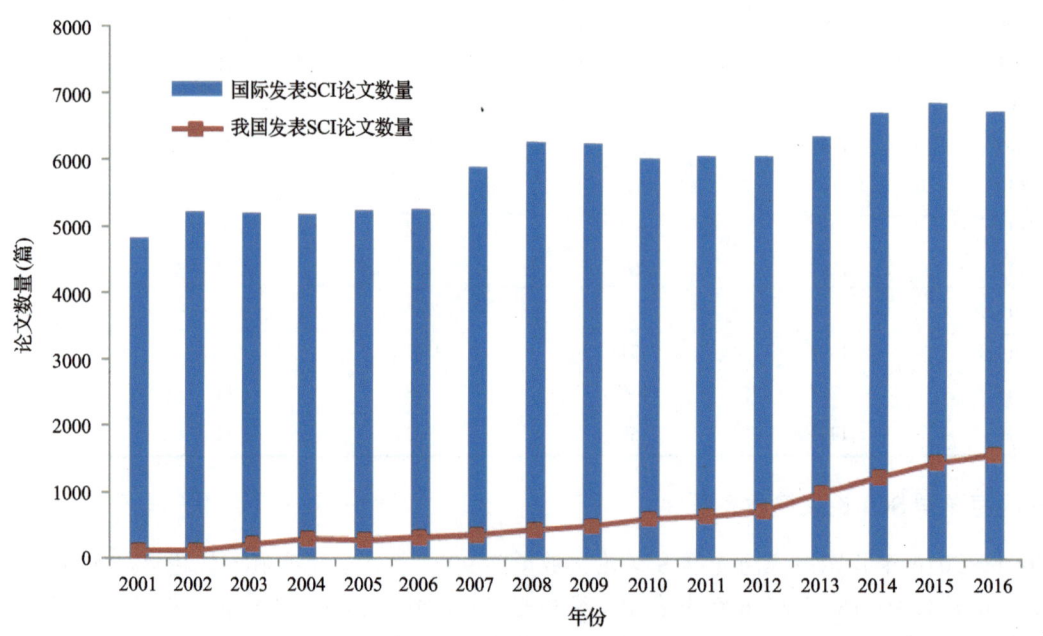

图 1-19　2001~2016 年我国与国际发表海洋学 SCI 论文数量变化趋势

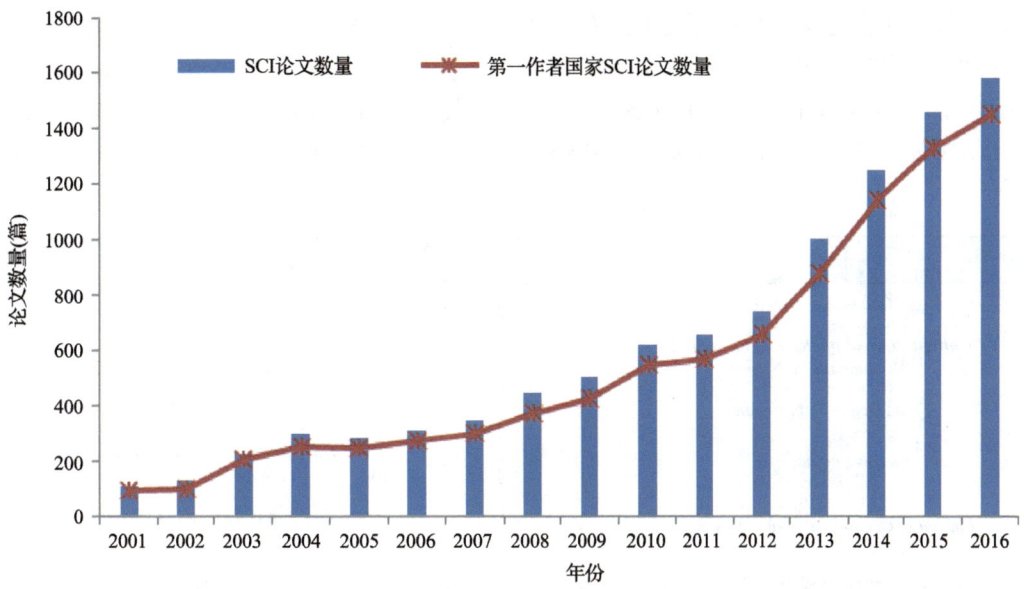

图 1-20 2001～2016 年我国海洋学 SCI 全部论文与第一作者国家署名 SCI 论文数量变化趋势

表 1-2 2001～2016 年我国海洋科技论文 SCIE 发文的学科分布

序号	WOS 学科分类	论文数量(篇)
1	海洋学	10 389
2	海洋工程	4 634
3	土木工程	2 508
4	湖沼学	1 350
5	气象学与大气科学	1 220
6	水资源学	1 213
7	地学交叉科学	1 206
8	海洋与淡水生物学	1 094
9	机械工程	1 054
10	工程学交叉科学	750
11	生态学	233
12	地球化学与地球物理学	200
13	地质工程	157
14	采矿与选矿	157
15	渔业学	143
16	化学交叉科学	141
17	电子与电气工程	115
18	遥感	62
19	环境科学	45
20	力学	45
21	古生物学	43
22	能源和燃料	3

海洋领域 SCI 论文期刊分布如表 1-3 所示，统计了 2001～2016 年我国海洋科技领域发表 SCI 论文前 20 的期刊。其中，载文量在 1000 篇以上的期刊为 Acta Oceanologica Sinica、Chinese Journal of Oceanology and Limnology，其次为 China Ocean Engineering、Ocean Engineering 和 Journal of Ocean University of China 和 Journal of Geophysical Research oceans 等期刊，其发文数量在 500 篇以上。

表 1-3　2001～2016 年我国海洋科技 SCI 论文发表前 20 的期刊及其发文数量

序号	期刊名称	发文数量（篇）	序号	期刊名称	发文数量（篇）
1	Acta Oceanologica Sinica	1479	11	Marine Ecology Progress Series	159
2	Chinese Journal of Oceanology and Limnology	1173	12	Marine Georesources & Geotechnology	154
3	China Ocean Engineering	874	13	Journal of Marine Systems	138
4	Ocean Engineering	709	14	Terrestrial Atmospheric and Oceanic Sciences	138
5	Journal of Ocean University of China	554	15	Journal of Atmospheric and Oceanic Technology	133
6	Journal of Geophysical Research-Oceans	518	16	Marine Geology	125
7	Estuarine Coastal and Shelf Science	320	17	Deep-Sea Research Part II-Topical Studies in Oceanography	123
8	Continental Shelf Research	294	18	Journal of Oceanography	122
9	Journal of Navigation	186	19	Ocean & Coastal Management	117
10	Applied Ocean Research	185	20	Marine Chemistry	112

2001～2016 年我国海洋科技 SCI 论文的主要发文机构中排名前 19 位的机构如表 1-4 所示。其中，中国科学院排名第一，论文数量为第二名中国海洋大学的 1.5 倍以上，发表论文在 1000 篇以上的机构还有排名第三的原国家海洋局[①]。其后依次为大连理工大学、厦门大学、上海交通大学、华东师范大学、浙江大学、河海大学及中国水产科学研究院等机构。

表 1-4　2001～2016 年我国发表海洋科技 SCI 论文排名前 19 的发文机构及其发表论文数量

序号	机构名称/中文	论文数量（篇）
1	中国科学院	2709
2	中国海洋大学	1783
3	原国家海洋局	1258
4	大连理工大学	588
5	厦门大学	518
6	上海交通大学	457
7	华东师范大学	318
8	浙江大学	316
9	河海大学	287
10	中国水产科学研究院	248
11	香港科技大学	219
12	天津大学	218

① 2018 年 3 月，国务院机构改革，组建自然资源部，整合国家海洋局职责，不再保留国家海洋局，对外保留国家海洋局牌子。

续表

序号	机构名称/中文	论文数量(篇)
13	同济大学	200
14	上海海洋大学	199
15	哈尔滨工程大学	197
16	南京大学	165
17	中山大学	114
18	香港大学	137
19	南京信息工程大学	130

三、海洋科技著作出版种类明显增长

2002~2016 年，我国海洋科研机构的海洋科技著作出版种类总体呈现增长态势(图 1-21)，年均增长率为 13.97%。其中，2002~2005 年海洋科技著作出版种类处于稳定增长阶段，年均增长率为 17.27%；2006~2007 年与 2008~2009 年海洋科技著作出版种类快速增长，增长率分别为 104.41% 与 64.47%；2010~2016 年海洋科技著作出版种类年均增长率为 10.64%。

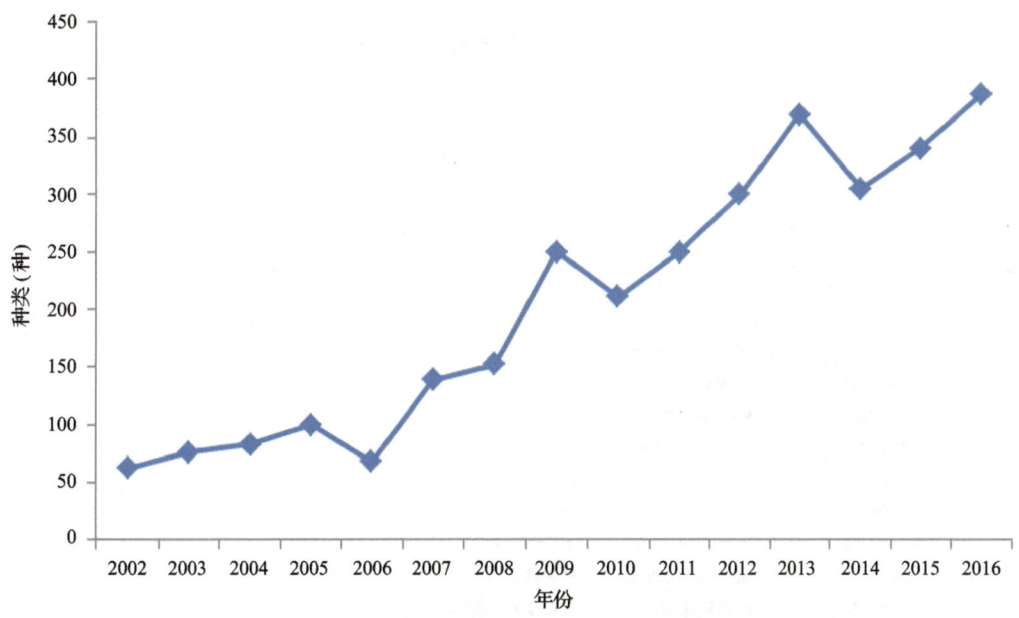

图 1-21　2002~2016 年我国海洋科技著作出版种类变化趋势

四、海洋领域工程索引(EI)论文先增后降

工程索引(Engineering Index，EI)是美国工程师学会联合会创办的一种关于工程技术领域文献的综合性情报检索工具，是工程技术界认可的非常重要的检索工具。其收录了近 2000 万条数据，收录范围涉及 76 个国家、190 多个工程学科、3600 余种期刊、80 多个图书连续出版物、8 万余个会议录及 8 万余篇学位论文，还有上百种贸易杂志等。本报告对 EI 数据库中海洋领域相关论文进行统计分析，了解全球和中国在海洋领域的研究发展态势。2001~2016 年，全球海洋学领域相关文献共 275 611 条，中国相关文献有 55 876 条。如图 1-22 所示，2001~2012 年，我国海洋领域 EI 论文数量及占全球 EI 论文数量的比重稳定上升。近 15 年来我国 EI 论文数量增长速度远远超过全球 EI 论文数量增

长速度。2012年中国海洋EI论文数量是2001年的32倍,占全球比重从2001年的4.74%上升到2016年的25%以上。

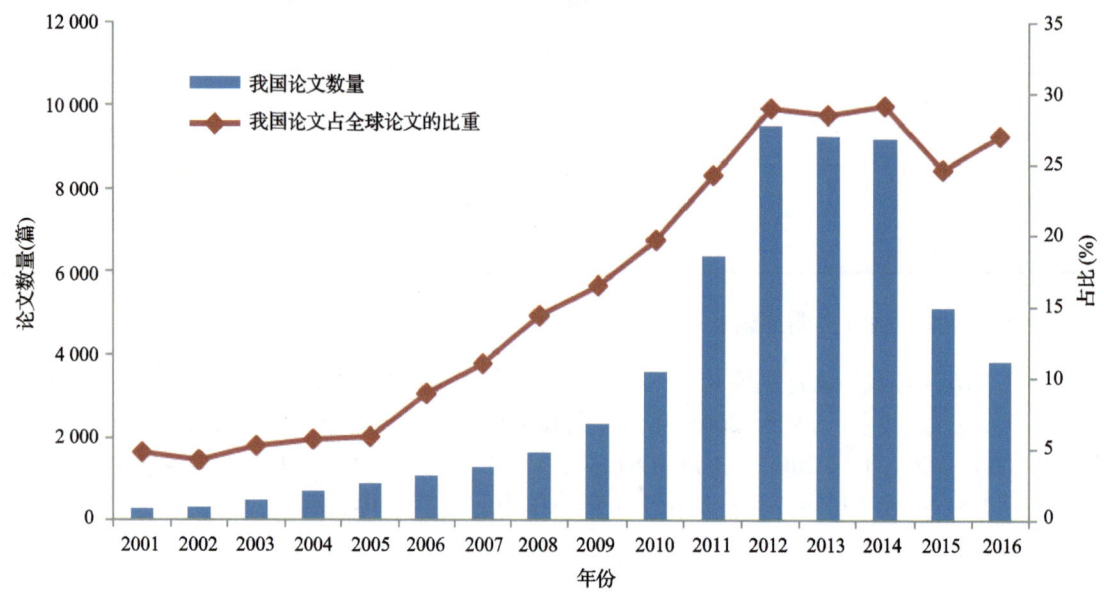

图1-22 2001～2016年我国海洋领域EI论文数量及占全球比重变化趋势

我国海洋领域EI论文的学科分布与国际相似,但我国在舰艇领域的论文占比更大,此外,与化学相关的主题(如一般化工产品、有机化合物等)分布也较多,如图1-23所示。

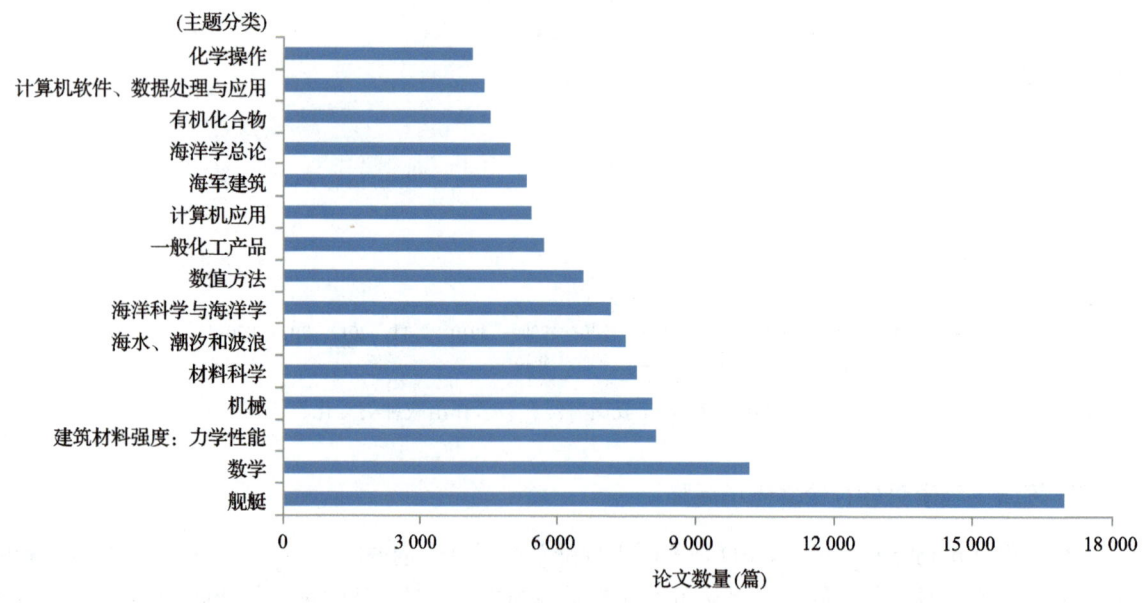

图1-23 我国海洋领域EI论文产出学科分布

中国发表海洋领域EI论文最多的15个机构如图1-24所示。中国涉海研究机构众多,中国科学院在海洋领域的科技论文数占全国的份额为8.32%。

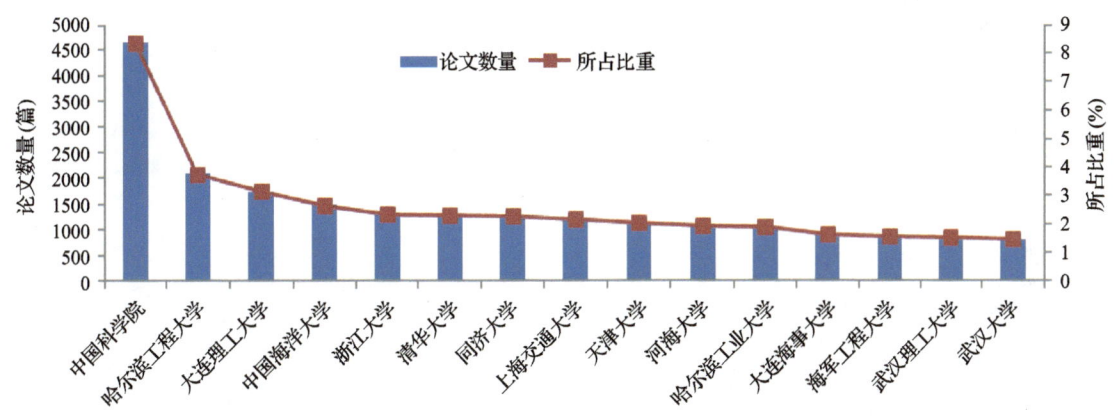

图 1-24 中国发表海洋领域 EI 论文最多的 15 个机构的论文数量及其占全国的比重

五、海洋领域专利申请量涨势强劲

2001～2016 年，我国海洋领域专利申请数量逐年增长，年均增长率为 23.90%，自 2006 年以来增长显著，2012～2016 年专利年申请数量维持在 4000 件以上，如图 1-25 和图 1-26 所示。2006 年中国海洋专利迎来了飞速发展，2010～2016 年每年增长 700 件以上。由于专利数据存在滞后性，近 3 年数据仅供参考，但仍可看出我国目前海洋领域技术处于高速发展期。

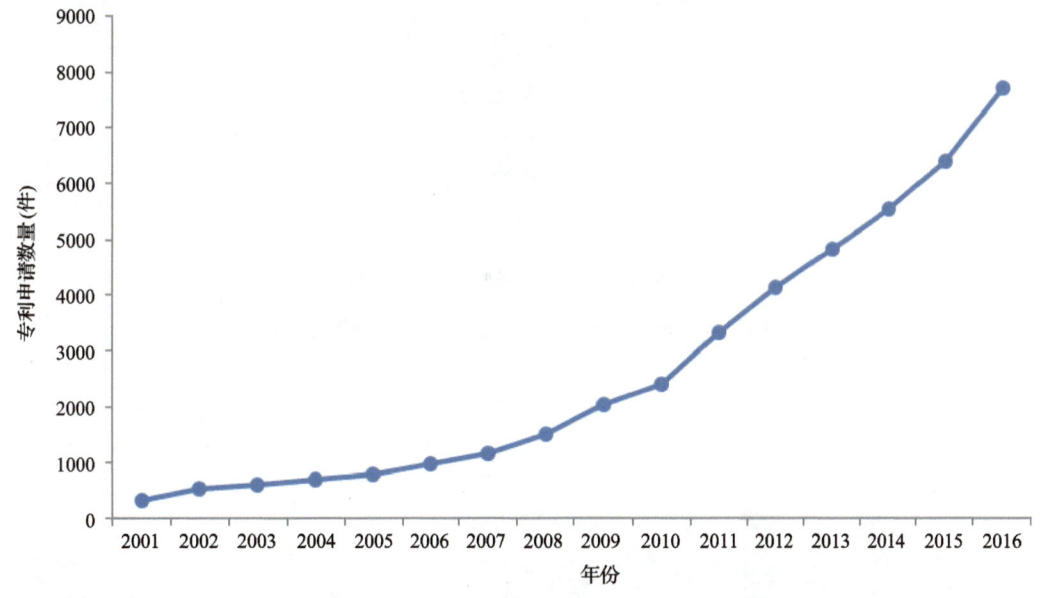

图 1-25 2001～2016 年我国海洋领域专利申请数量趋势

我国海洋领域专利类型中，发明专利占 62%（图 1-27），说明我国海洋专利目前技术研发居多，创新潜力较大。

我国海洋领域专利申请机构的前 15 位（图 1-28）中，企业有 4 个，均是海洋石油领域企业。高等学校有 8 家，主要分布在山东、浙江和上海等地。科研院所有 3 个，主要是中国科学院及中国水产科学研究院的相关海洋研究机构。

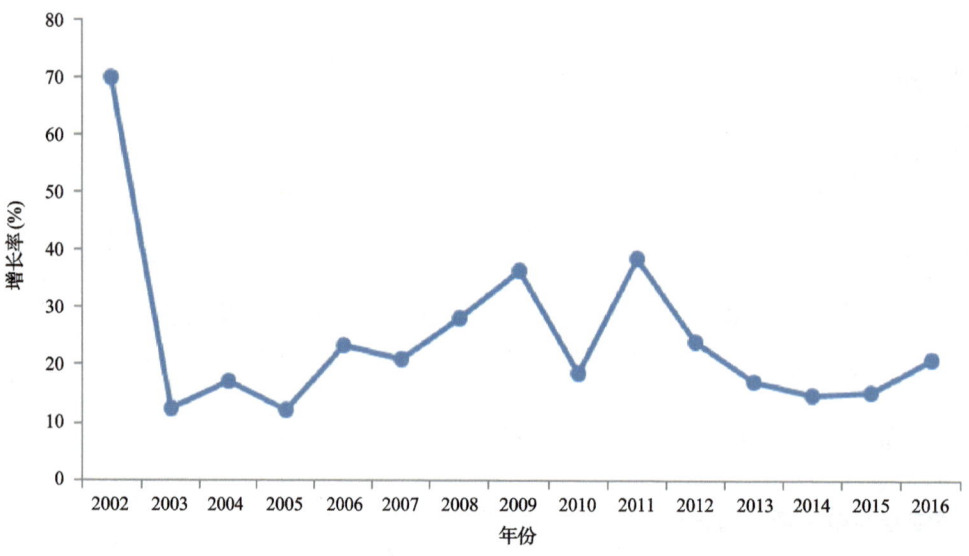

图 1-26　2002～2016 年我国海洋领域专利申请数量年度增长率变化

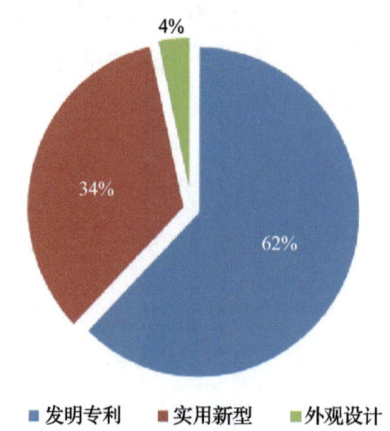

图 1-27　2001～2016 年我国海洋领域专利类型比例

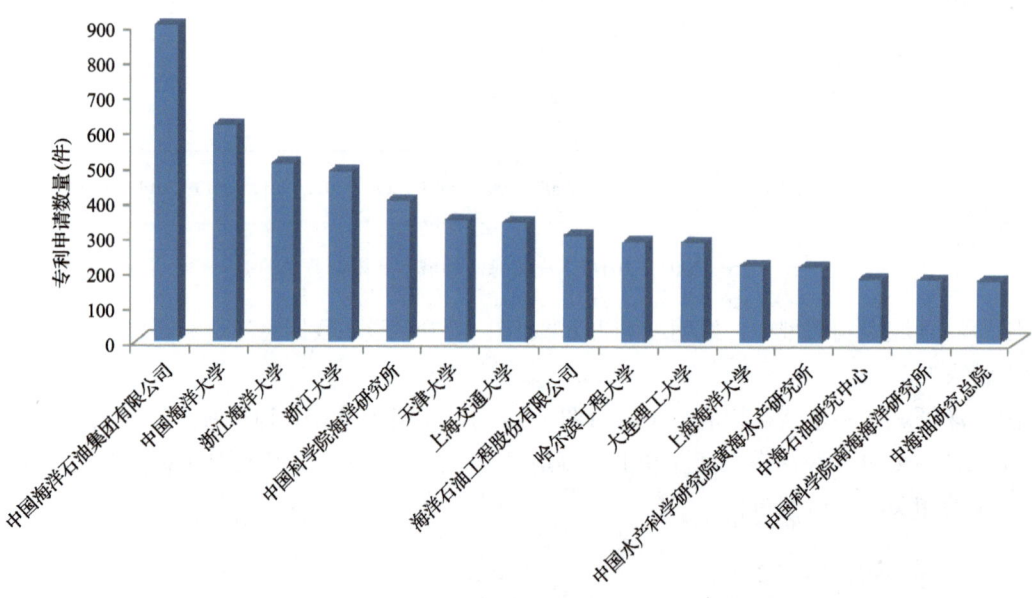

图 1-28　2001～2016 年我国海洋领域专利申请机构的前 15 位及其专利申请数量

我国海洋领域前 15 位专利申请机构的专利类型（图 1-29）中，以发明专利为主，占比约为 70%，且高等学校和科研院所的发明专利比例明显高于企业；外观设计专利仅有 2 件，中国海洋石油集团有限公司[①]和哈尔滨工程大学各 1 件。

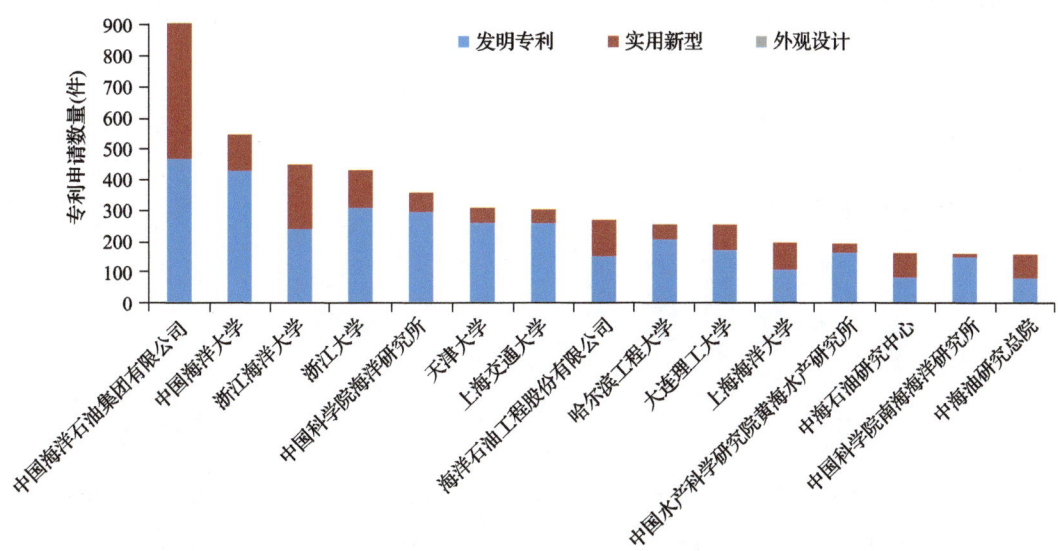

图 1-29　2001～2015 年我国海洋领域前 15 位专利申请机构专利类型

"外观设计"由于数据过小，图中显示不明显

2001～2006 年，我国海洋领域专利主要申请省（直辖市）中，山东省因其较多的涉海科研机构与大学而居首位，江苏省、浙江省分列第二位、第三位，北京市位列第四。其他沿海省（直辖市）中，福建省专利申请数量相对较少（图 1-30）。

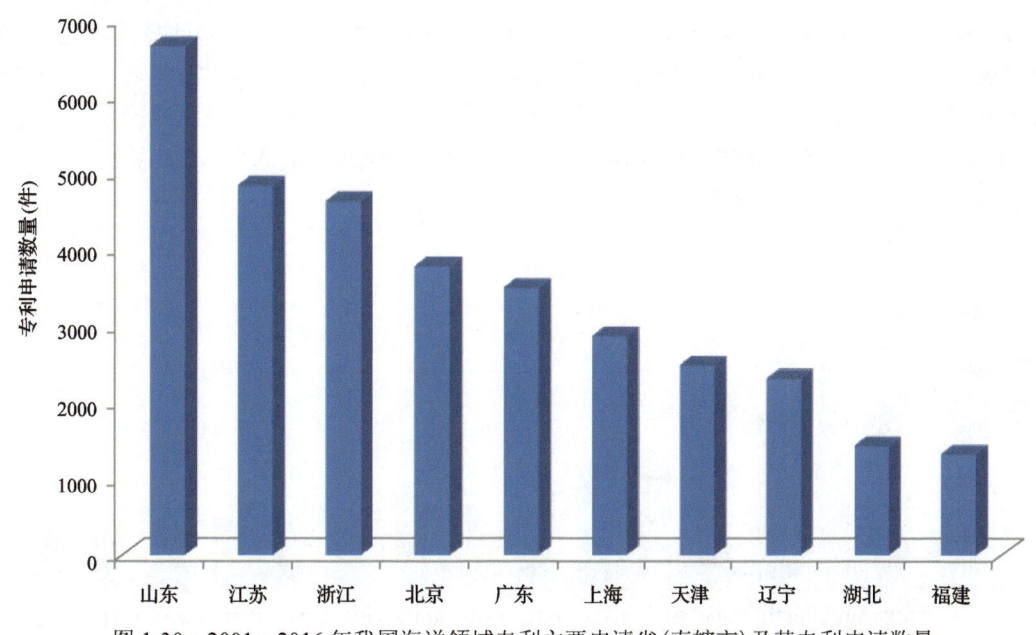

图 1-30　2001～2016 年我国海洋领域专利主要申请省（直辖市）及其专利申请数量

从主要申请城市来看，青岛专利申请数量高于其他城市，大连、杭州、广州专利申请数量相当，

---

① 2017 年 11 月 1 日，中国海洋石油总公司正式更名为中国海洋石油集团有限公司

武汉和南京作为非沿海城市表现突出，如图1-31所示。

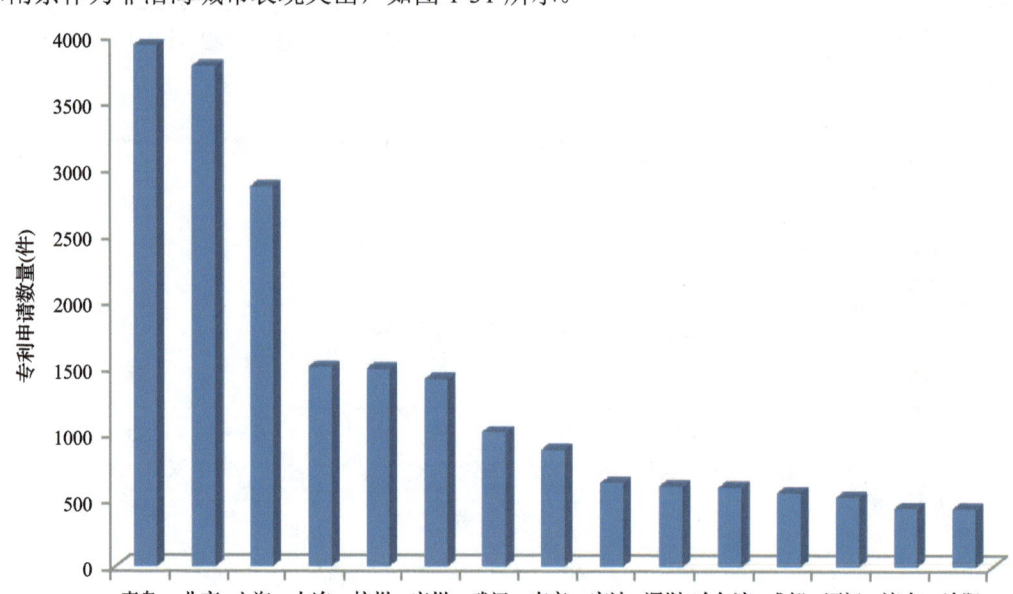

图 1-31　2001～2016 年我国海洋领域专利主要申请城市及其专利申请数量

2001～2016 年我国海洋领域出现频次较高的 15 类专利依次为：C02F（污水、污泥污染处理）、A01K（畜牧业；禽类、鱼类、昆虫的管理；捕鱼；饲养或养殖其他类不包含的动物；动物的新品种）、B63B（船舶或其他水上船只；船用设备）、G01N（借助测定材料的化学或者物理性质来测试或分析材料）、F03B（液力机械或液力发动机）、E02B（水利工程）、B01D（分离）、E21B（土层或岩石的钻进）、A61K（医学用配置品）、E02D（基础、挖方、填方、地下或水下结构物）、A23L（不包含在 A21D 或 A23B～A23J 小类中的食品、食料或非酒精饮料）、C12N（微生物或酶）、C09D（涂料组合物，如色漆、清漆或天然漆；填充浆料；化学涂料或油墨的去除剂；油墨；改正液；木材着色剂；用于着色或印刷的浆料或固体；原料为此的应用）、F16L（管子；管接头或管件；管子、电缆或护管的支撑；一般的绝热方法）、H01B（电缆；导体；绝缘体；导电、绝缘或介电材料的选择），如图 1-32 所示。

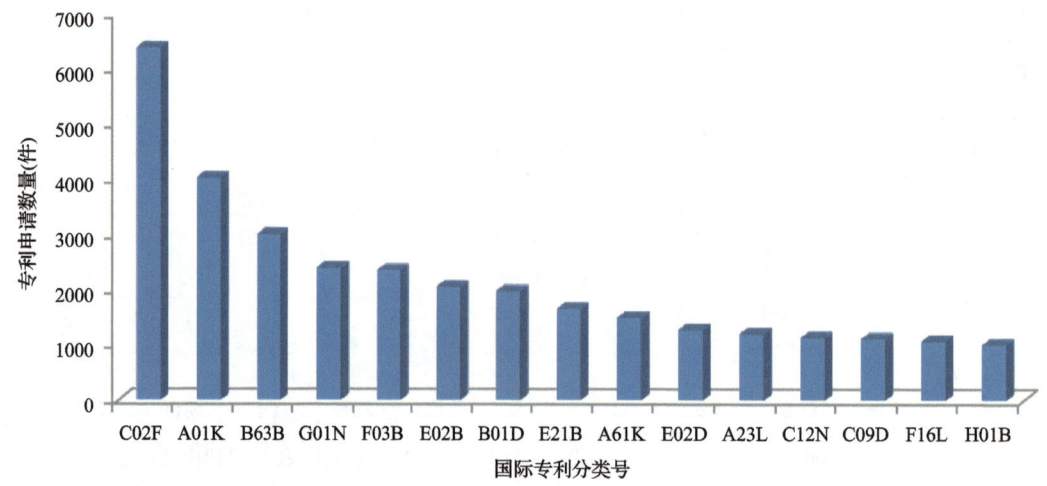

图 1-32　2001～2016 年我国海洋领域主要国际专利分类号（IPC 号）的专利申请数量

第五节 高等学校海洋创新稳步提升

高等学校对国家创新发展具有举足轻重的作用。近年来，我国高等学校的海洋创新资源投入和海洋创新成果产出逐渐增加，海洋创新发展态势良好。需要说明的是，本部分数据是以主要涉海高等学校和涉海学科为依据提取的，按照其涉海比例系数加权求和所得（主要涉海高等学校及其涉海比例系数和涉海学科及其涉海比例系数分别见附录八和附录九）。

一、高等学校海洋创新人力资源结构逐渐优化

高等学校教学与科研人员是指高等学校在册职工在统计年度内，从事大专以上教学、研究与发展、研究与发展成果应用及科技服务工作人员，以及直接为上述工作服务的人员，包括统计年度内从事科研活动累计工作时间一个月以上的外籍和高等教育系统以外的专家和访问学者。如图1-33所示，2009~2016年我国涉海高等学校教学与科研人员总体呈上升趋势，从2015年起有所下降，但总体波动程度不大。其中，科学家与工程师、高级职称人员数量总体呈增长态势，科学家与工程师占教学与科研人员的比例略有波动；高级职称人员占教学与科研人员的比例由37.50%上升至43.10%。

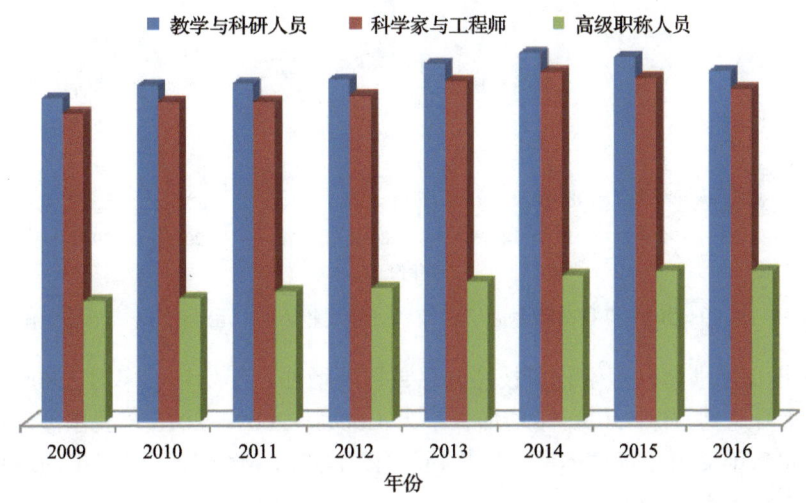

图1-33 2009~2016年我国涉海高等学校教学与科研人员和其中的科学家与工程师、高级职称人员数量（人）变化趋势

高等学校研究与发展人员是指统计年度内，从事研究与发展工作时间占本人教学、科研总时间10%以上的教学与科研人员。2009~2016年我国涉海高等学校研究与发展人员基本稳定（图1-34）。其中，科学家与工程师、高级职称人员数量总体呈增长态势，科学家与工程师占研究与发展人员的比例略波动下降，由95.99%下降到95.47%，高级职称人员占研究与发展人员的比例略有波动。

二、高等学校海洋创新投入逐渐增加

2009~2016年，我国涉海高等学校科技经费投入总体上呈增加趋势，年均增长率达11.38%；政府资金投入呈增长态势，年均增长率达12.12%；我国涉海高等学校的当年内部支出大幅增长（图1-35），2016年内部支出是2009年的66.98倍。

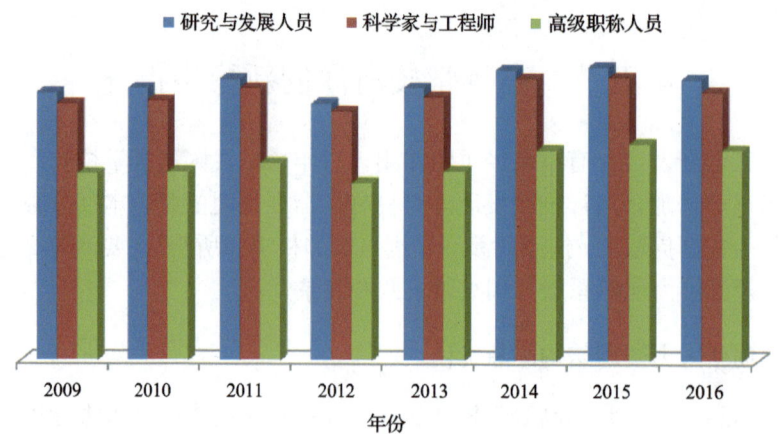

图 1-34 2009~2016 年我国涉海高等学校研究与发展人员和其中的科学家与工程师、高级职称人员数量(人)变化趋势

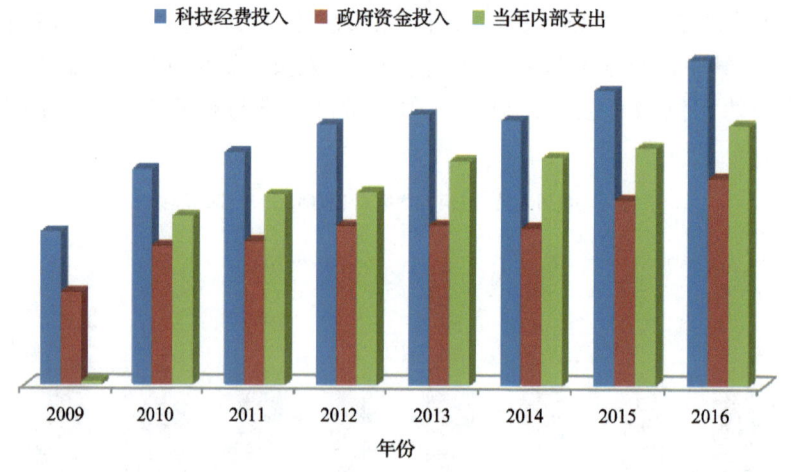

图 1-35 2009~2016 年我国涉海高等学校科技经费投入(千元)与支出(千元)变化趋势

2009~2016 年我国涉海高等学校科技课题总数逐渐增加，年均增长率为 5.80%；科技课题当年投入人数总体呈上升趋势(图 1-36)，年均增长率为 1.00%。2009~2016 年我国涉海高等学校科技课题当年投入经费和当年支出经费总体呈现增长趋势(图 1-37)，当年投入经费年均增长率达 10.35%，当年支出经费年均增长率达 11.06%。

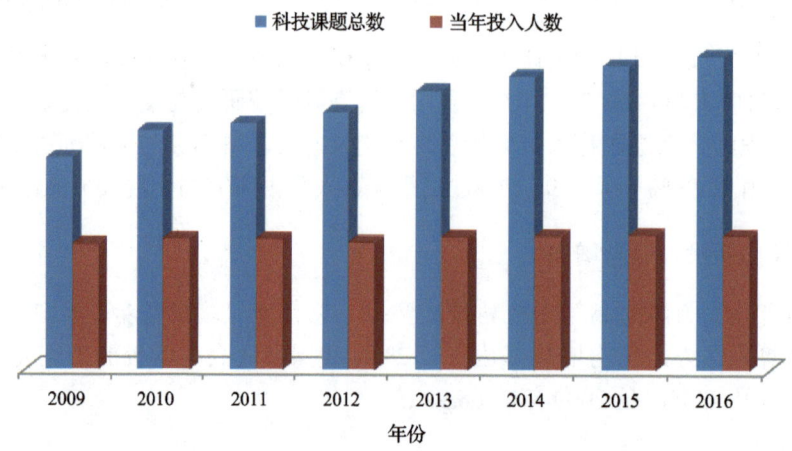

图 1-36 2009~2016 年我国涉海高等学校科技课题总数(项)和当年投入人数(人)变化趋势

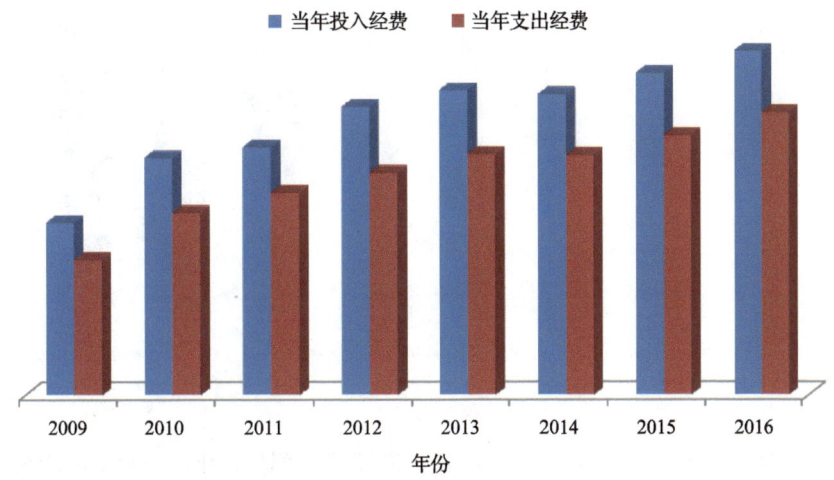

图 1-37 2009~2016 年我国涉海高等学校科技课题当年投入经费(千元)和当年支出经费(千元)变化趋势

三、高等学校海洋创新产出逐渐增加

2009~2016 年,我国涉海高等学校科技成果中发表的学术论文数量总体呈现增长趋势,年均增长率为 5.10%。其中,在国外学术刊物发表的学术论文数量增长更为明显,年均增长率为 13.27%(图 1-38)。技术转让签订的合同数在 2009~2010 年增长最为迅猛(图 1-39),增长率达到 67.60%,2010~2011 年有所下降,之后开始逐年增长,年均增长率为 15.24%。

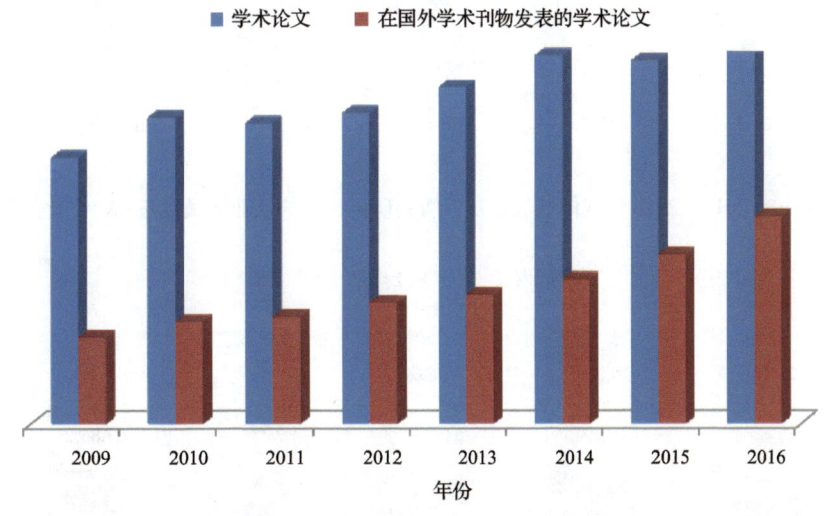

图 1-38 2009~2016 年我国涉海高等学校科技成果中发表学术论文数量(篇)变化趋势

四、高等学校涉海科研机构稳定发展

2012~2016 年我国高等学校涉海科研机构中的从业人员逐步增加(图 1-40),其中,博士毕业和硕士毕业人员数量也呈增长态势。同时,从业人员中博士毕业人员占比由 51.76% 上升到 54.40%,硕士毕业人员占比由 27.56% 上升到 27.86%(图 1-41)。

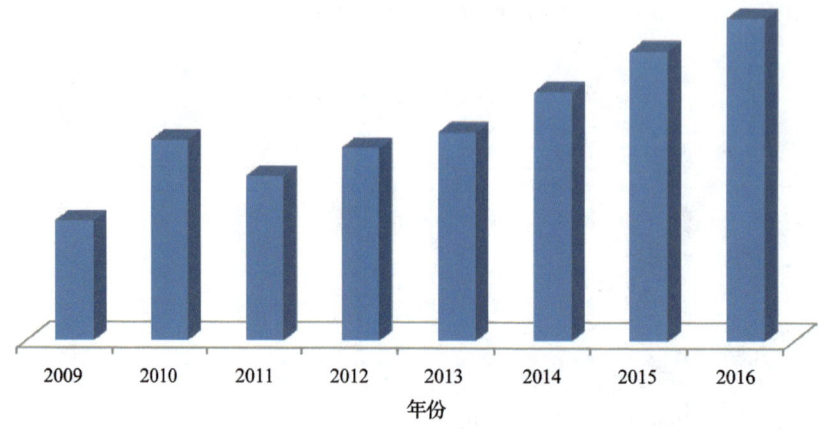

图 1-39　2009～2016 年我国涉海高等学校技术转让签订的合同数(项)变化趋势

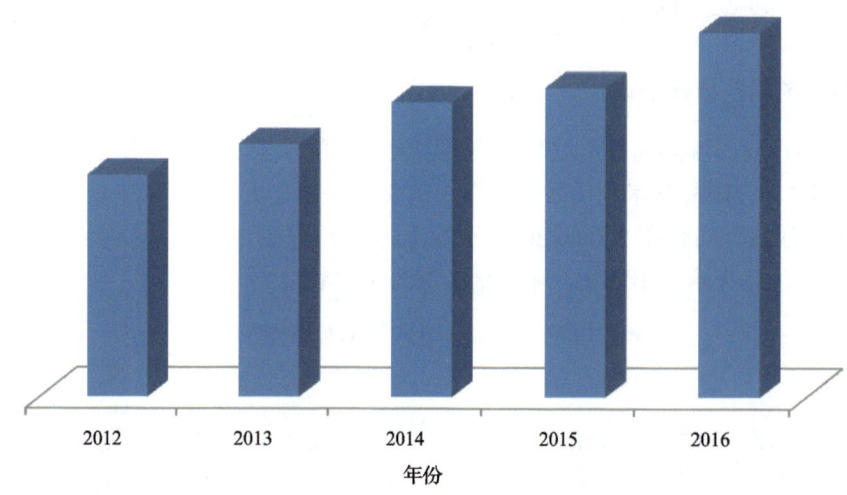

图 1-40　2012～2016 年我国高等学校涉海科研机构中的从业人员数量(人)变化趋势

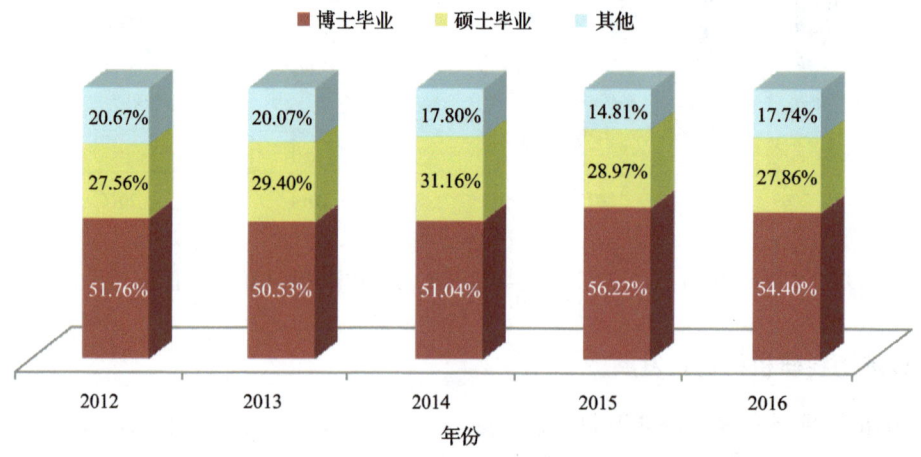

图 1-41　2012～2016 年我国高等学校涉海科研机构中的从业人员学历结构

2012～2016 年我国高等学校涉海科研机构中的科技活动人员数量总体呈增长趋势(图 1-42)。其中，高级职称人员占比由 60.00%下降至 59.01%，中级职称人员占比由 28.46%上升至 30.86%，初级职称人员占比由 7.76%下降至 6.30%(图 1-43)。

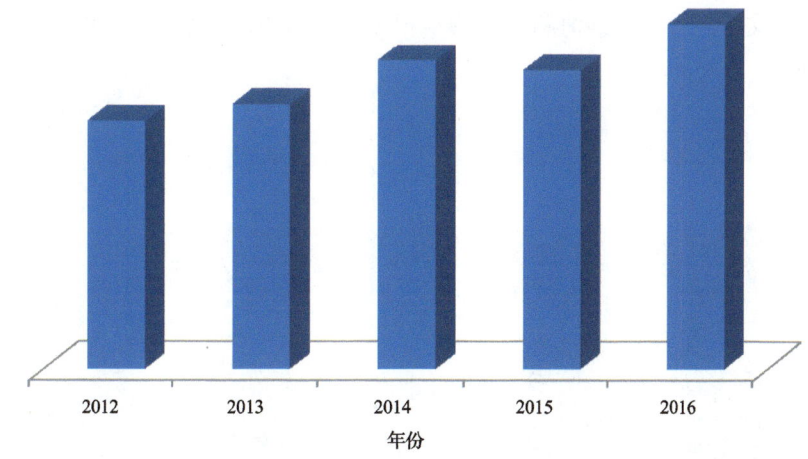

图 1-42　2012~2016 年我国高等学校涉海科研机构中的科技活动人员数量(人)变化趋势

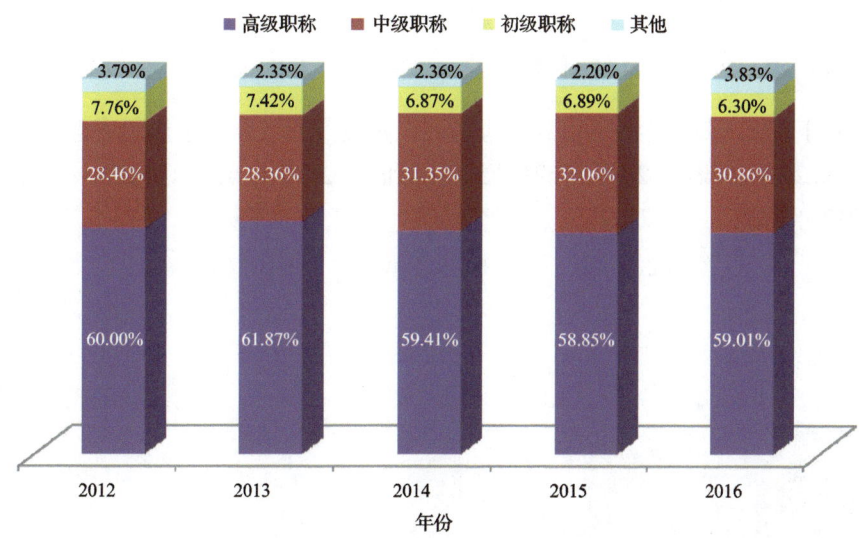

图 1-43　2012~2016 年我国高等学校涉海科研机构中的科技活动人员职称结构

2012~2016 年我国高等学校涉海科研机构的科技经费支出不断增加(图 1-44),2016 年的当年经费内部支出比 2012 年增加 88.14%,2016 年的 R&D 经费支出比 2012 年增加 108.02%。

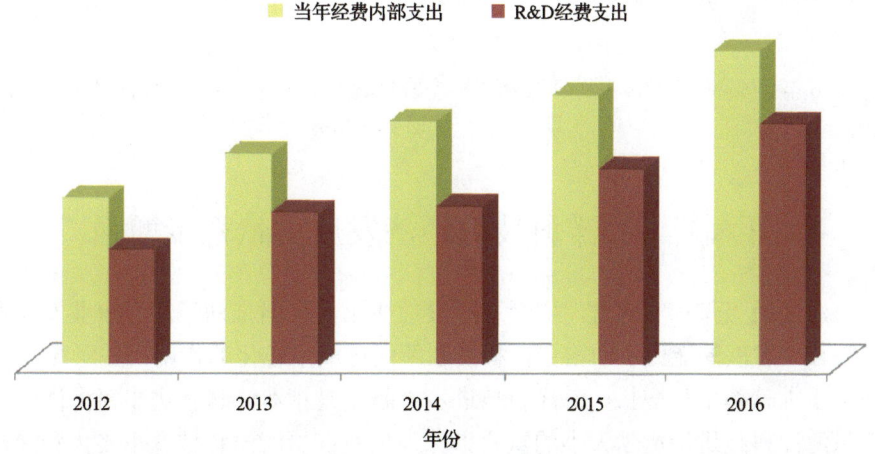

图 1-44　2012~2016 年我国高等学校涉海科研机构科技经费支出(千元)变化趋势

2012～2016年我国高等学校涉海科研机构承担项目总数逐渐增加(图1-45)，2016年与2012年相比增加54.00%。

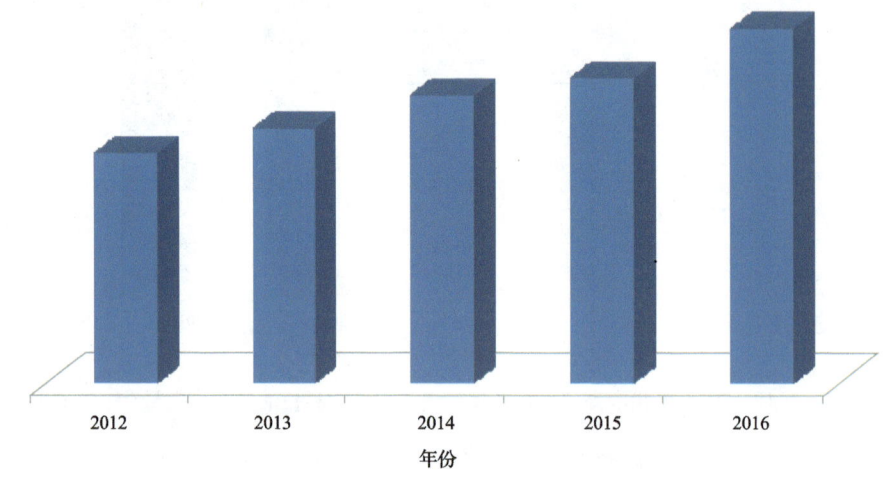

图1-45　2012～2016年我国高等学校涉海科研机构承担项目数量(项)变化趋势

2012～2016年，我国高等学校涉海科研机构的固定资产原值保持增加趋势(图1-46)，2016年比2012年增加了52.25%。其中，2016年的仪器设备原值比2012年增加了51.26%，2016年进口仪器设备原值比2012年增加了70.52%。

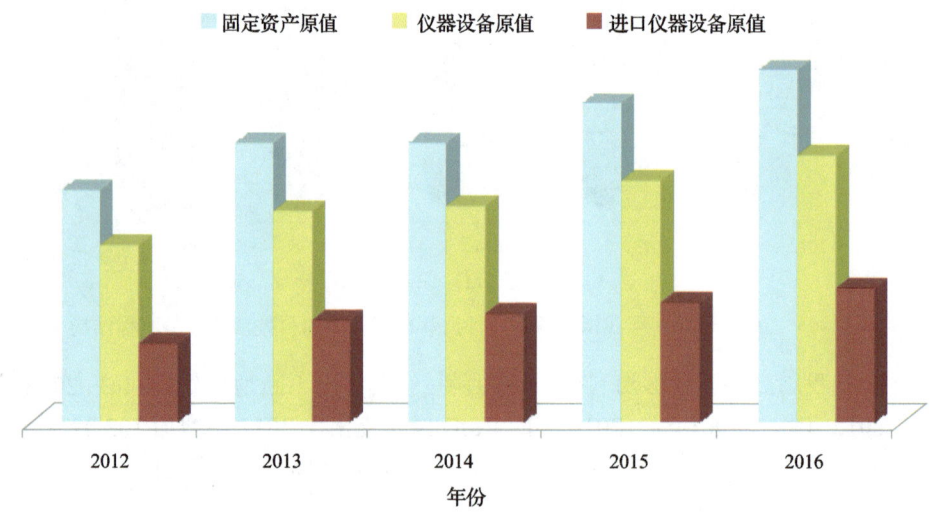

图1-46　2012～2016年我国高等学校涉海科研机构的固定资产原值(千元)和其中的仪器设备原值(千元)、进口仪器设备原值(千元)变化趋势

第六节　海洋科技对经济发展贡献稳步增强

近年来，海洋创新工作扎实推进，取得了阶段性成果，全面推动了海洋事业发展进程。海洋科技服务海洋经济社会发展的能力不断增强，科技创新促进成果转化的作用日益彰显。

海洋科技进步贡献率平稳增长。海洋科技进步贡献率是指海洋科技进步对海洋经济增长的贡献份额，它是度量海洋科技进步贡献大小的重要指标，也是衡量海洋科技竞争实力和海洋科技转化为现实生产力水平的综合性指标。《"十三五"国家科技创新规划》在发展目标中明确提出"科技创新作为经济工作的重要方面,在促进经济平衡性、包容性和可持续性发展中的作用更加突出,科技进步贡献率达到60%"。根据历年《中国海洋统计年鉴》数据,基于加权改进的索洛余值法(测算过程见附录五),测算我国"十一五"期间(2006~2010年)、"十二五"期间(2011~2015年),以及"十一五"以来直至"十三五"开局之年(2006~2016年)的海洋科技进步贡献率(表1-5)。

表1-5 我国海洋科技进步贡献率(%)

年份	产出增长率	资本增长率	劳动增长率	海洋科技进步贡献率
2006~2010	12.86	10.10	4.05	54.4
2011~2015	10.97	6.74	2.72	64.2
2006~2016	10.53	6.00	2.56	65.9

从表1-5可以看出,"十一五"期间我国海洋科技进步贡献率为54.4%,2011~2015年达到64.2%,2006~2016年继续提高到65.9%。也就是说,在2006~2016年我国海洋生产总值的年均增长率12.67%中,有65.9%来自海洋科技进步的贡献,高于《全国科技兴海规划(2016—2020年)》提出的目标,2016年作为国家"十三五"开局之年,为"十三五"期间海洋创新发展开创了新局面。

海洋科技成果转化能力发展良好。海洋科技成果转化率是指进行自我转化或转化生产,处于投入应用或生产状态,并达到成熟应用的海洋科技成果占全部海洋科技应用成果的百分率。海洋科技成果能否迅速而有效地转化为现实生产力,是一个国家海洋事业发展和腾飞的关键。加快海洋科技成果向现实生产力转化,促进新产品、新技术的更新换代和推广应用,是海洋科技进步工作的中心环节,也是促进海洋经济发展由粗放型向集约型转变的关键所在。《全国海洋经济发展"十三五"规划》提出2020年海洋科技成果转化率达到55%以上。根据科学技术部海洋科技统计和海洋科技成果登记数据,2000~2016年海洋科技成果转化率达到50.0%(测算过程见附录六),海洋科技成果转化能力仍有较大提升空间。

第二章　国家海洋创新指数评价

国家海洋创新指数是一个综合指数，由海洋创新资源、海洋知识创造、海洋创新绩效和海洋创新环境4个分指数构成。考虑海洋创新活动的全面性和代表性，以及基础数据的可获取性，本报告选取20个指标（指标体系见附录一），反映海洋创新的质量、效率和能力。

国家海洋创新指数显著上升，海洋创新能力大幅提高。设定2004年我国的国家海洋创新指数基数值为100，则2016年国家海洋创新指数为240，2004～2016年国家海洋创新指数的年均增长率为7.58%，"十二五"期间年均增长率为6.63%，保持平稳发展态势。

海洋创新资源分指数总体呈上升趋势，2004～2016年年均增长率为8.00%。其中，"研究与发展经费投入强度"和"研究与发展人力投入强度"两个指标的年均增长率分别为10.84%和11.17%，是拉动海洋创新资源分指数上升的主要力量。

海洋知识创造分指数增长强劲，2004～2016年年均增长率达到10.84%。"本年出版科技著作"和"万名R&D人员的发明专利授权数"两个指标增长较快，年均增长率分别达13.69%和14.22%，高于其他指标值，成为推动海洋知识创造的主导力量。

海洋创新绩效分指数在4个分指数中增长较慢，2004～2016年年均增长率仅为4.94%。"海洋劳动生产率"在海洋创新绩效分指数的6个指标中增长较为稳定，年均增长率为10.95%，对海洋创新绩效的增长起着积极的推动作用。

海洋创新环境分指数呈上升趋势，2004～2016年年均增长率为5.39%，尤其在2005～2009年快速增长，这得益于指标"沿海地区人均海洋生产总值"与"R&D经费中设备购置费所占比重"的迅速增长。

第一节 海洋创新指数综合评价

一、国家海洋创新指数总体上升

将2004年我国的国家海洋创新指数定为基数100，则2016年国家海洋创新指数达到240（图2-1），2004~2016年，年均增长率为7.58%。

图2-1 国家海洋创新指数历年变化及增长率

2004~2016年国家海洋创新指数总体呈上升趋势，增长率出现不同程度的波动，最为突出的是2007年，国家海洋创新指数由2006年的110增长为2007年的140，增长率达到峰值27.27%。主要原因包括：国际金融危机下，我国采取了有效应对措施；随着国家对海洋创新投入逐渐加大，效果开始显现；越来越多的科研机构走向海洋。以2009年为界，2004~2009年，国家海洋创新指数上升趋势较快，年均增长率为10.78%；而2010~2016年，增长有所减缓，年均增长率为6.12%。

二、国家海洋创新指数与4个分指数关系密切

4个分指数对国家海洋创新指数的影响各不相同，呈现不同程度的上升态势（表2-1，图2-2）。海洋创新资源分指数与国家海洋创新指数变化趋势较为接近，各年均呈正向增长。海洋知识创造分指数得分值明显高于国家海洋创新指数，说明海洋知识创造分指数对国家海洋创新指数增长有较大的正贡献。海洋创新绩效分指数基本呈现平稳缓慢的线性增长，年均增长率出现小范围波动，与国家海洋创新指数的年均增长率有所差异。海洋创新环境分指数，在2004~2006年与国家海洋创新指数分值及其变化趋势最为接近，2007~2016年其值比国家海洋创新指数要低，但是变化趋势仍较接近。

2004~2016年，我国海洋创新资源分指数年均增长率为8.00%（表2-2），各年均呈现正增长，充分体现了我国海洋创新资源投入持续增加的发展态势。

表 2-1　国家海洋创新指数和各分指数变化

年份	综合指数	分指数			
	国家海洋创新指数 A	海洋创新资源 B_1	海洋知识创造 B_2	海洋创新绩效 B_3	海洋创新环境 B_4
2004	100	100	100	100	100
2005	106	102	111	103	106
2006	110	105	109	112	113
2007	140	162	152	115	130
2008	148	172	164	125	132
2009	167	197	197	127	146
2010	168	199	195	136	144
2011	178	208	214	146	146
2012	196	221	251	153	158
2013	215	236	306	157	162
2014	216	239	288	164	174
2015	231	246	327	169	181
2016	240	252	344	178	188

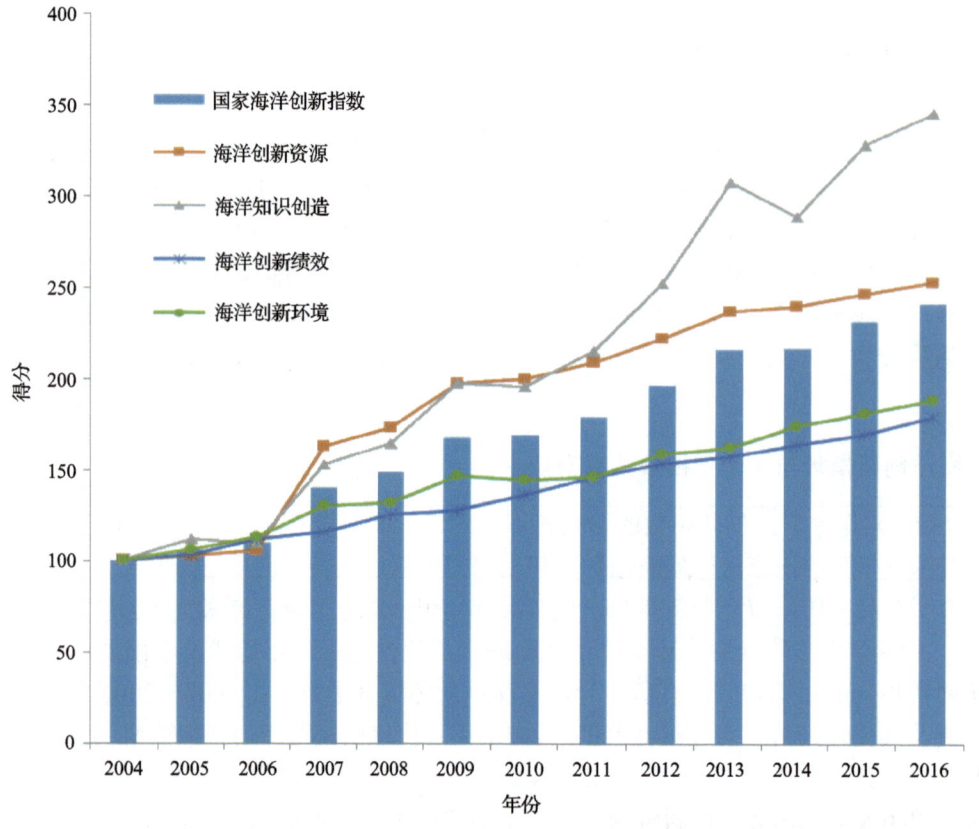

图 2-2　2004~2016 年国家海洋创新指数及其分指数得分变化趋势

表 2-2 国家海洋创新指数和分指数增长率(%)

年份	综合指数	分指数			
	国家海洋创新指数 A	海洋创新资源 B_1	海洋知识创造 B_2	海洋创新绩效 B_3	海洋创新环境 B_4
2004	—	—	—	—	—
2005	5.61	2.18	11.48	3.22	5.58
2006	3.87	3.04	−1.99	8.05	6.79
2007	27.52	53.90	39.36	3.49	15.17
2008	5.93	6.41	7.49	8.45	1.28
2009	12.59	14.30	20.32	1.71	11.08
2010	0.94	0.90	−1.07	6.70	−1.30
2011	5.93	4.65	9.91	7.20	1.10
2012	9.82	6.18	17.36	5.04	8.69
2013	9.91	6.92	21.94	2.60	2.08
2014	0.28	1.05	−6.09	4.20	7.40
2015	6.79	2.94	13.65	3.37	3.92
2016	4.27	2.47	5.14	5.50	3.98
年均增长率	7.58	8.00	10.84	4.94	5.39

2004～2016 年,海洋知识创造分指数对我国海洋创新能力大幅提升的贡献较大,年均增长率达到 10.84%(图 2-3)。表明我国海洋科研能力迅速增强,海洋知识创造及其转化运用为海洋创新活动提供了强有力的支撑。海洋知识创造能力的提高为增强国家原始创新能力、提高自主创新水平提供了重要支撑。

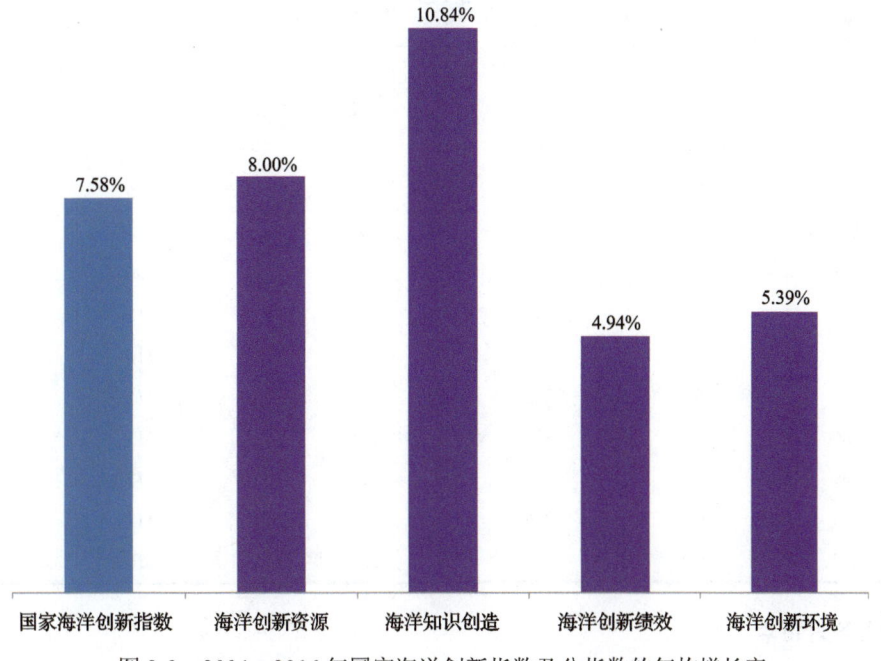

图 2-3 2004～2016 年国家海洋创新指数及分指数的年均增长率

促进海洋经济发展是海洋创新活动的最终目标,是进行海洋创新能力评价不可或缺的组成部分。从近年来的变化趋势来看,我国海洋创新绩效稳步提升。2004～2016 年我国海洋创新绩效分指数年均增长率达到4.94%,各年均呈现正增长态势,增长率最高值出现在2008年,为8.45%(表2-2)。

海洋创新环境是海洋创新活动顺利开展的重要保障。我国海洋创新的总体环境极大改善,2004~2009年海洋创新环境分指数一直呈上升趋势(表2-1),年均增长率为7.87%,但在2010年首次出现了负增长,2004~2016年年均增长率为5.39%,在4个分指数中位列第三(图2-3)。

第二节 海洋创新资源分指数评价

海洋创新资源能够反映一个国家对海洋创新活动的投入力度。创新型人才资源供给能力及创新所依赖的基础设施投入水平,是国家持续开展海洋创新活动的基本保障。海洋创新资源分指数采用如下5个指标:①研究与发展经费投入强度;②研究与发展人力投入强度;③R&D人员中博士占比;④科技活动人员占海洋科研机构从业人员的比重;⑤万名科研人员承担的课题数。通过以上指标,从资金投入、人力投入等角度对我国海洋创新资源投入和配置能力进行评价。

一、海洋创新资源分指数升势趋稳

2016年海洋创新资源分指数得分为252(表2-3),比2015年略微上升,2004~2016年的年均增长率为8.00%。从历史变化情况来看,2007年和2009年海洋创新资源分指数的涨幅最为明显,年增长率分别为53.90%与14.30%;2010年以后,海洋创新资源分指数逐年增长。

表2-3 海洋创新资源分指数及其指标得分

年份	分指数	指标				
	海洋创新资源	研究与发展经费投入强度	研究与发展人力投入强度	R&D人员中博士占比	科技活动人员占海洋科研机构从业人员的比重	万名科研人员承担的课题数
	B_1	C_1	C_2	C_3	C_4	C_5
2004	100	100	100	100	100	100
2005	102	94	95	112	101	110
2006	105	99	94	120	102	112
2007	162	204	194	174	105	134
2008	172	216	202	198	108	138
2009	197	285	248	207	109	136
2010	199	250	246	238	111	149
2011	208	270	267	251	110	143
2012	221	287	272	286	111	149
2013	236	333	317	281	113	137
2014	239	337	322	275	117	143
2015	246	362	316	289	117	145
2016	252	344	356	286	116	157

二、指标变化各有特点

从海洋创新资源的5个指标得分的变化趋势(图2-4)来看,有3个指标整体呈快速上升趋势,2个指标波动平稳。其中,"研究与发展经费投入强度"指标波动幅度最大,其次是"研究与发展人力投入强度"指标和"R&D人员中博士占比"指标,2004~2016年,3个指标整体均呈现增长态势,年均增长率分别为 10.84%、11.17% 和 9.15%，是拉动海洋创新资源分指数整体上升的主要力量。

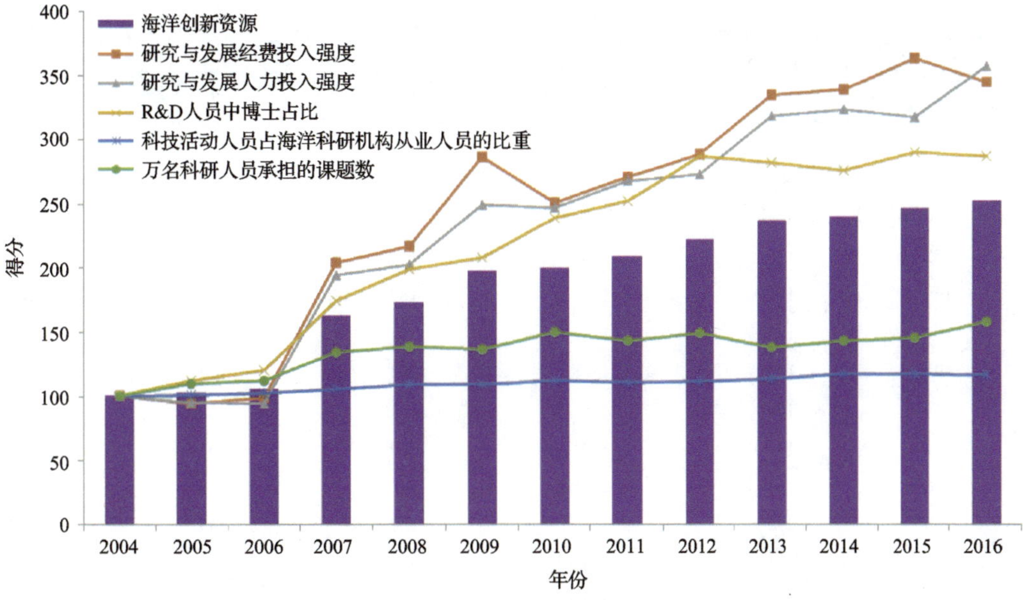

图 2-4　海洋创新资源分指数及其指标得分变化趋势

"科技活动人员占海洋科研机构从业人员的比重"指标能够反映一个国家海洋创新活动科研力量的强度，2004～2016 年增长率基本持平，年均增长率为 1.24%，趋于平稳。

"万名科研人员承担的课题数"指标能够反映海洋科研人员从事海洋创新活动的强度。其变化趋势以 2009 年为界，2004～2008 年呈稳定上涨趋势，年均增长率为 8.44%；2009 年出现负增长，之后不断波动；2010 年、2012 年与 2016 年得分较高，2010～2016 年，该指标年均增长率为 0.88%。

第三节　海洋知识创造分指数评价

海洋知识创造是创新活动的直接产出，能够反映一个国家海洋领域的科研产出能力和知识传播能力。海洋知识创造分指数选取如下 5 个指标：①亿美元经济产出的发明专利申请数；②万名 R&D 人员的发明专利授权数；③本年出版科技著作；④万名科研人员发表的科技论文数；⑤国外发表的论文数占总论文数的比重。通过以上指标论证我国海洋知识创造的能力和水平，既能反映科技成果产出效应，又综合考虑了发明专利、科技论文、科技著作等各种成果产出。

一、海洋知识创造分指数波动上升

从海洋知识创造分指数及其增长率来看（表 2-4，图 2-5），我国的海洋知识创造分指数总体呈波动上升趋势，从 2004 年的 100 增长至 2016 年的 344，年均增长率达 10.84%。从图 2-5 可看出，海洋知识创造分指数增长大致划分为两个阶段。第一个阶段是 2013 年之前，海洋知识创造呈现相对缓慢的上升趋势，年均增长率为 13.25%；第二个阶段是 2013 年以后，海洋知识创造分指数不断波动且呈稳定趋势，2013～2016 年的年均增长率达到 3.92%。分指数在 2016 年得分最高，这主要得益于指标"万名 R&D 人员的发明专利授权数"，该指标极大地提高了海洋知识创造分指数指标得分。

表 2-4　海洋知识创造分指数及其指标得分

年份	分指数 海洋知识创造 B_2	指标				
		亿美元经济产出的 发明专利申请数 C_6	万名 R&D 人员的 发明专利授权数 C_7	本年出版科技著作 C_8	万名科研人员发表的 科技论文数 C_9	国外发表的论文数占 总论文数的比重 C_{10}
2004	100	100	100	100	100	100
2005	111	85	124	120	117	110
2006	109	69	142	82	125	129
2007	152	116	118	167	167	193
2008	164	141	145	182	164	187
2009	197	174	172	301	161	177
2010	195	165	178	254	157	220
2011	214	160	238	301	152	219
2012	251	183	319	370	161	224
2013	306	352	298	445	149	290
2014	288	298	327	367	143	304
2015	327	274	475	440	136	311
2016	344	299	493	466	146	315

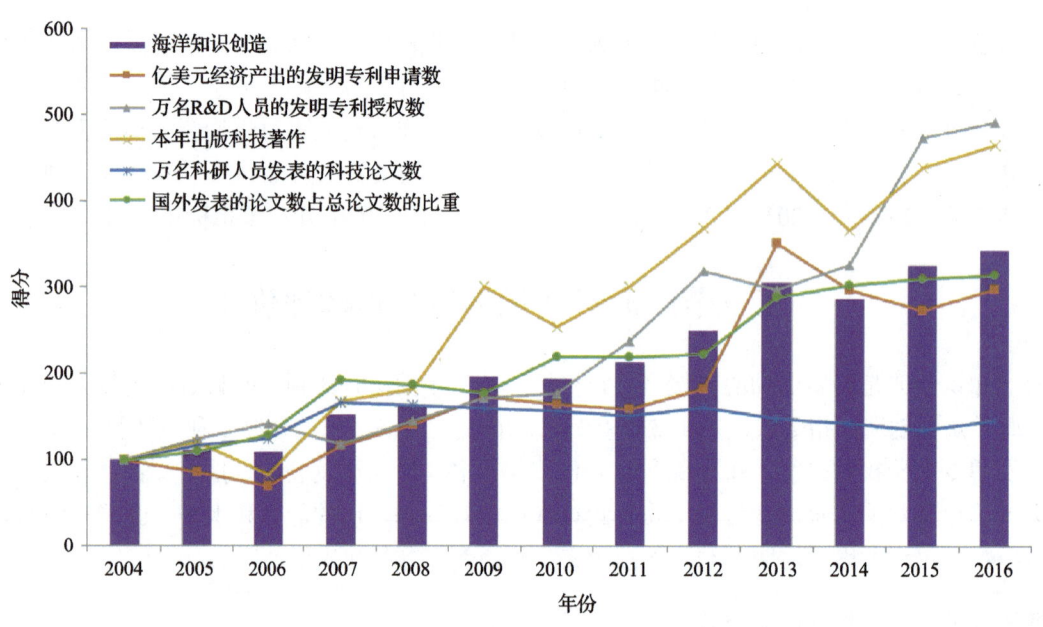

图 2-5　海洋知识创造分指数及其指标得分变化趋势

二、各指标贡献不一

从海洋知识创造 5 个指标的变化趋势来看(图 2-5),"本年出版科技著作"和"万名 R&D 人员的发明专利授权数"两个指标波动幅度最大,"本年出版科技著作"在 2010~2013 年增长迅猛,由 254 上升到 445,年均增长率为 20.48%。"万名 R&D 人员的发明专利授权数"在 2014~2015 年迅速增长,由 327 上升到 475,增长率为 45.26%,其他年份两个指标呈现小幅波动。总体来看,2004~2016 年,两个指标呈现波动增长趋势,年均增长率分别达 13.69%和 14.22%。

2004～2016 年,"亿美元经济产出的发明专利申请数"指标呈现波动增长态势,年均增长率为 9.55%。其中,2004～2006 年,该指标均是负增长;2007 年后进入快速上升阶段,2013 年达到峰值 352。2013～2015 年该指标有所下降,2016 年得分为 299。

"万名科研人员发表的科技论文数"即平均每万名科研人员发表的科技论文数,反映了科学研究的产出效率。"国外发表的论文数占总论文数的比重"是指一国发表的科技论文中国外发表论文的比重,反映了科技论文的对外普及程度。2004～2016 年,2 个指标得分增长相对缓慢,年均增长率分别为 3.22%和 10.04%。

第四节　海洋创新绩效分指数评价

海洋创新绩效能够反映一个国家开展海洋创新活动所产生的效果和影响。海洋创新绩效分指数选取如下 6 个指标:①海洋科技成果转化率;②海洋科技进步贡献率;③海洋劳动生产率;④科研教育管理服务业占海洋生产总值的比重;⑤单位能耗的海洋经济产出;⑥海洋生产总值占国内生产总值的比重。以上指标反映了我国海洋创新活动所带来的效果和影响。

一、海洋创新绩效分指数有序上升

表 2-5 是海洋创新绩效分指数及其指标的历年得分。从分指数得分情况来看,我国的海洋创新绩效分指数从 2004 年的 100 增长至 2016 年的 178,呈现平稳的增长态势,年均增长率为 4.94%,在 4 个分指数中增长最为缓慢。

表 2-5　海洋创新绩效分指数及其指标得分

年份	分指数	指标					
	海洋创新绩效 B_3	海洋科技成果转化率 C_{11}	海洋科技进步贡献率 C_{12}	海洋劳动生产率 C_{13}	科研教育管理服务业占海洋生产总值的比重 C_{14}	单位能耗的海洋经济产出 C_{15}	海洋生产总值占国内生产总值的比重 C_{16}
2004	100	100	100	100	100	100	100
2005	103	105	92	113	96	109	104
2006	112	109	108	130	92	122	109
2007	115	112	106	145	91	133	105
2008	125	115	127	165	92	148	103
2009	127	118	119	176	94	153	103
2010	136	120	114	211	85	177	107
2011	146	122	134	238	84	190	105
2012	153	124	143	258	86	201	105
2013	157	126	137	276	87	211	104
2014	164	128	144	305	92	207	104
2015	169	130	145	322	96	219	103
2016	178	129	149	348	106	235	104

二、指标变化趋势稳定

"海洋科技成果转化率"是衡量海洋科技转化为现实生产力水平的重要指标。2004～2016 年我国海洋科技成果转化率呈现上升趋势,年均增长率为 2.13%。总体来看,2010 年以前我国海洋科技成果转化率增长缓慢,2010 年以后趋于稳定(图 2-6)。

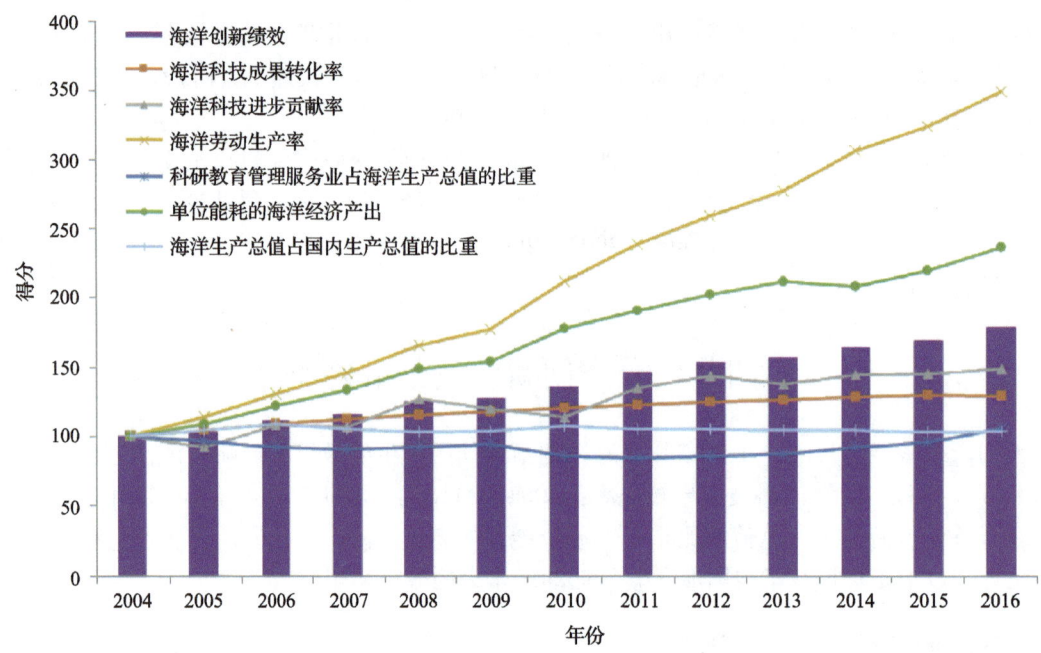

图 2-6　海洋创新绩效分指数及其指标得分变化趋势

2004～2016 年，"海洋科技进步贡献率"指标总体波动不大，稳中有升。

"海洋劳动生产率"是指海洋科技人员的人均海洋生产总值，反映海洋创新活动对海洋经济产出的作用。2004～2016 年，"海洋劳动生产率"指标迅速增长，年均增长率为 10.95%，是创新绩效分指数 6 个指标中增长最快最稳定的指标(图 2-6)。

"科研教育管理服务业占海洋生产总值的比重"能够反映海洋科研、教育、管理及服务等活动对海洋经济的贡献程度，该指标 2004～2016 年年均增长率为 0.45%，表明海洋科研、教育、管理及服务等活动对海洋经济的贡献程度呈现相对上升的趋势。

"单位能耗的海洋经济产出"采用万吨标准煤能源消耗的海洋生产总值，测度海洋创新对减少资源消耗的作用，也反映出一个国家海洋经济增长的集约化水平。2004～2016 年，"单位能耗的海洋经济产出"指标增长迅速，年均增长率为 7.40%，呈现稳定的增长态势。

"海洋生产总值占国内生产总值的比重"反映海洋经济对国民经济的贡献，用来测度海洋创新对海洋经济的推动作用。该指标变化不明显，2016 年其得分仅比 2004 年增长 4 分，增长速度缓慢，2004～2016 年的年均增长率为 0.31%。

第五节　海洋创新环境分指数评价

海洋创新环境包括创新过程中的硬环境和软环境，是提升我国海洋创新能力的重要基础和保障。海洋创新环境分指数反映一个国家海洋创新活动所依赖的外部环境，主要是制度创新和环境创新。海洋创新环境分指数选取如下 4 个指标：①沿海地区人均海洋生产总值；②R&D 经费中设备购置费所占比重；③海洋科研机构科技经费筹集额中政府资金所占比重；④R&D 人员人均折合全时工作量。

一、海洋创新环境明显改善

2004～2016 年，海洋创新环境分指数总体上呈现稳步增长态势(表 2-6，图 2-7)，由 2004 年的 100 上升至 2016 年的 188，年均增长率达到 5.39%，其中 2007 年增长率为 15.04%，达到峰值，这主要得益于指标"R&D 经费中设备购置费所占比重"的迅速增长，指标得分由 2006 年的 105 增长至 2007 年的 151。

表 2-6　海洋创新环境分指数及其指标历年得分

年份	分指数	指标			
	海洋创新环境 B_4	沿海地区人均海洋生产总值 C_{17}	R&D 经费中设备购置费所占比重 C_{18}	海洋科研机构科技经费筹集额中政府资金所占比重 C_{19}	R&D 人员人均折合全时工作量 C_{20}
2004	100	100	100	100	100
2005	106	120	100	99	103
2006	113	143	105	96	107
2007	130	170	151	107	92
2008	132	194	131	107	94
2009	146	209	181	102	93
2010	144	253	128	100	96
2011	146	289	104	99	92
2012	158	326	107	104	97
2013	162	352	94	103	98
2014	174	391	96	106	102
2015	181	414	97	108	103
2016	188	448	99	107	97

图 2-7　海洋创新环境分指数及其指标得分变化趋势

二、优势指标与劣势指标并存

海洋创新环境分指数的指标中,一直保持上升趋势的指标是"沿海地区人均海洋生产总值",年均增长率为13.31%,该指标与海洋创新环境分指数的得分和走势都最为接近,在4个指标中增长最快。

相对劣势指标为"R&D经费中设备购置费所占比重"和"R&D人员人均折合全时工作量"。"R&D经费中设备购置费所占比重"得分有一定的波动,总体呈下滑趋势,最高值出现在2009年,之后逐渐下降,由2009年的181下降至2016年的99。"R&D人员人均折合全时工作量"得分整体呈现波动下滑趋势,得分由2004年的100降至2016年的97。

第三章 区域海洋创新指数评价

区域海洋创新是国家海洋创新的重要组成部分,深刻影响着国家海洋创新的格局。本报告分析了区域海洋创新的发展现状和特点,为我国海洋创新格局的优化提供了科技支撑和决策依据。

《推动共建丝绸之路经济带和21世纪海上丝绸之路的愿景与行动》中提出要"利用长三角、珠三角、海峡西岸、环渤海等经济区开放程度高、经济实力强、辐射带动作用大的优势"。从"一带一路"发展思路和我国沿海区域发展角度分析,我国沿海地区应积极优化海洋经济总体布局,实行优势互补、联合开发,充分发挥环渤海经济区、长江三角洲经济区、海峡西岸经济区、珠江三角洲经济区和环北部湾经济区五大经济区(海洋经济区的界定见附录七)的引领作用,推进形成我国北部、东部和南部三大海洋经济圈(海洋经济圈的界定见附录七)。

从我国沿海省(自治区、直辖市)[①]的区域海洋创新指数(区域海洋创新指数评价方法和指标体系说明见附录四)来看,2016年,我国11个沿海省(自治区、直辖市)可分为四个梯次,第一梯次为上海、广东;第二梯次包括山东、天津;第三梯次为江苏、福建和辽宁;第四梯次为浙江、河北、海南和广西。

从五大经济区的区域海洋创新指数来看,2016年,区域海洋创新能力较强的地区为珠江三角洲经济区、长江三角洲经济区及环渤海经济区,这些地区均有区域创新中心,而且呈现多中心的发展格局。

从三大海洋经济圈的区域海洋创新指数来看,2016年,我国海洋经济圈呈现北部、东部强而南部较弱的特点。北部海洋经济圈和东部海洋经济圈的区域海洋创新指数较高,表现出很强的原始创新能力,充分显示出我国重要海洋人才集聚地和海洋经济产业重点发展区域的优势。

---

① 本次评价仅包括我国11个沿海省(自治区、直辖市),不涉及香港、澳门和台湾

第一节　从沿海省(自治区、直辖市)看我国区域海洋创新发展

一、区域海洋创新梯次分明

根据 2016 年区域海洋创新指数得分(表 3-1,图 3-1),可将我国 11 个沿海省(自治区、直辖市)划分为 4 个梯次。

表 3-1　2016 年沿海省(自治区、直辖市)区域海洋创新指数与分指数得分

沿海省(自治区、直辖市)	综合指数	分指数			
	区域海洋创新指数 a	海洋创新资源 b_1	海洋知识创造 b_2	海洋创新绩效 b_3	海洋创新环境 b_4
上海	65.06	65.69	47.49	90.91	56.14
广东	61.51	47.88	95.64	57.67	44.85
山东	56.50	45.41	56.42	52.90	71.25
天津	54.90	64.40	25.35	70.83	59.03
江苏	49.83	82.47	50.55	38.98	27.32
福建	46.69	34.02	31.42	57.95	63.37
辽宁	44.48	59.68	60.05	31.24	26.95
浙江	36.33	34.28	40.83	36.69	33.53
河北	30.13	44.22	29.76	15.04	31.50
海南	22.74	8.84	2.57	57.07	22.47
广西	19.71	8.27	13.11	8.36	49.08

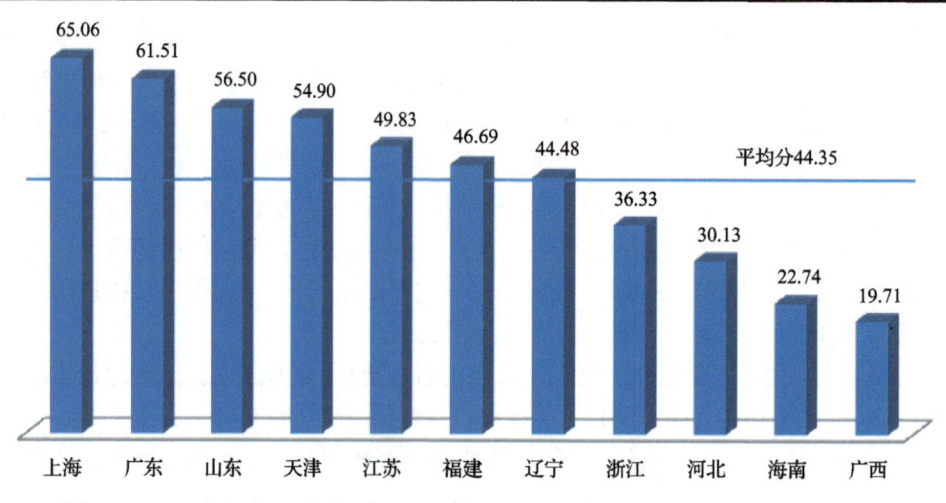

图 3-1　2016 年沿海 11 个省(自治区、直辖市)的区域海洋创新指数得分及平均分

从区域海洋创新指数来看,第一梯次为上海和广东,上海区域海洋创新指数得分为 65.06,相当于 11 个沿海省份平均水平的 1.47 倍,位居我国大陆 11 个沿海省(自治区、直辖市)首位,其海洋创新发展具备坚实的基础,表现出很强的原始创新能力;广东区域海洋创新指数得分为 61.51,排名由 2015 年的第三位上升至第二位,其海洋知识创造的快速发展拉动海洋创新能力的大幅提高。第二梯次包括山东和天津,其区域海洋创新指数得分分别为 56.50 和 54.90,高于 11 个沿海省(自治区、直辖市)的平均分 44.35。这两个地区有一定的海洋创新基础,长期以来积累了大量的创新资源,创新环境较好,创新绩效显著。第三梯次为江苏、福建和辽宁,其区域海洋创新指数得分分别为 49.83、

46.69 和 44.48，与平均分相近。这些地区近年来海洋经济发展较快，创新资源不断增多，创新环境明显改善，知识创造与创新绩效都进步较快。第四梯次为浙江、河北、海南和广西，其区域海洋创新指数得分分别为 36.33、30.13、22.74 和 19.71，低于国家的平均水平。从横向比较来看，浙江、河北、海南和广西海洋创新资源薄弱，知识创造效率不高，创新环境有待改善。

从海洋创新资源分指数来看，2016 年，海洋创新资源分指数得分超过平均分的沿海省（直辖市）有江苏、上海、天津、辽宁、广东和山东（图 3-2）。其中，江苏的区域海洋创新资源分指数得分为 82.47，远高于其他地区；上海、天津和辽宁的区域海洋创新资源分指数得分分别为 65.69、64.40 和 59.68，上海的该分指数得分虽然低于江苏，但其经费和人力的投入强度均位于 11 个沿海省（自治区、直辖市）首位。

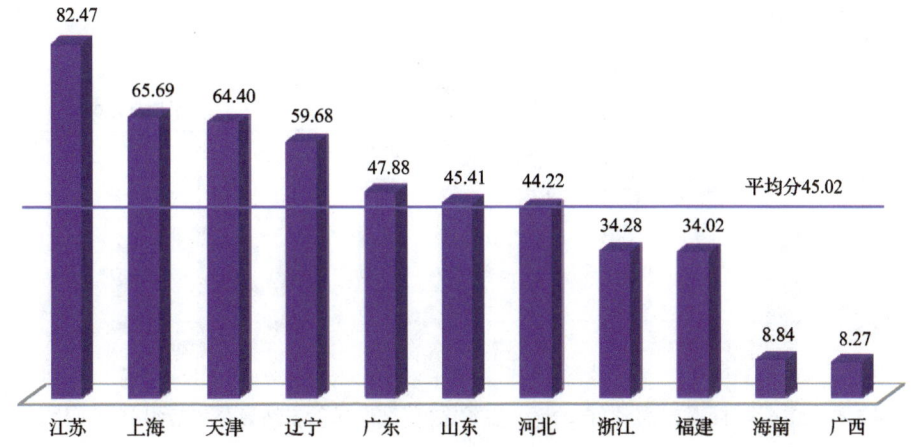

图 3-2　2016 年 11 个沿海省（自治区、直辖市）区域海洋创新资源分指数得分及平均分

从海洋知识创造分指数来看，2016 年，我国海洋知识创造分指数得分超过平均分的沿海省（直辖市）为广东、辽宁、山东、江苏和上海（图 3-3）。其中，广东的区域海洋知识创造分指数得分为 95.64，远高于 41.20 的平均分，较去年增长 56.58%，这与广东高产出、高质量的海洋科技著作和论文密不可分；辽宁的区域海洋知识创造分指数得分为 60.05，这主要得益于海洋科技发明专利；山东得分为 56.42，其中主要贡献来自于海洋科技著作和论文；江苏的区域海洋知识创造分指数得分为 50.55，其主要贡献来自于高产出、高质量的论文和专利；上海的区域海洋知识创造分指数得分为 47.49，这主要得益于海洋科技发明专利。

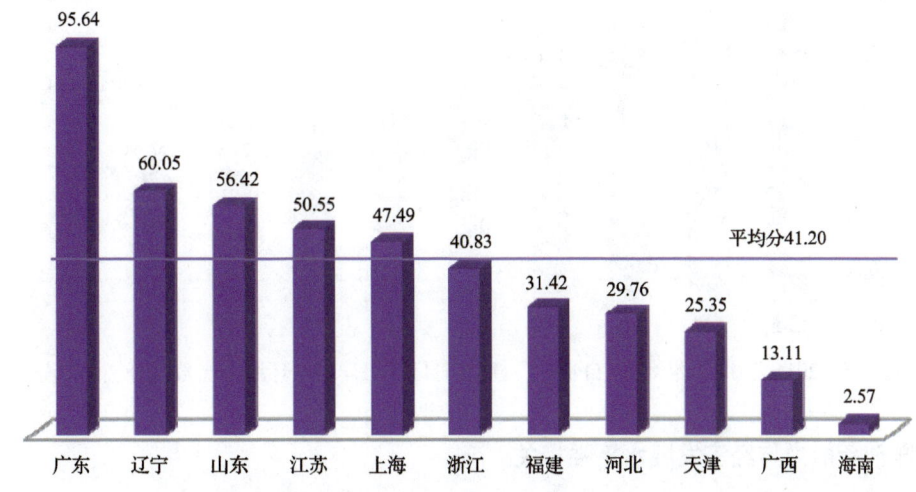

图 3-3　2016 年 11 个沿海省（自治区、直辖市）区域海洋知识创造分指数得分及平均分

从海洋创新绩效分指数来看，2016 年，海洋创新绩效分指数得分超过平均分的沿海省(直辖市)有上海、天津、福建、广东、海南和山东(图 3-4)。其中，上海的区域海洋创新绩效分指数得分为 90.91，主要原因在于其劳动生产率远高于其他地区，且拥有良好的海洋经济产出；天津的区域海洋创新绩效分指数得分为 70.83，紧随上海之后，这也得益于其较好的海洋经济产出；福建、广东、海南和山东的区域海洋创新绩效分指数得分分别为 57.95、57.67、57.07 和 52.90，海洋创新绩效各方面良好，整体处于 11 个沿海省(自治区、直辖市)平均水平之上。

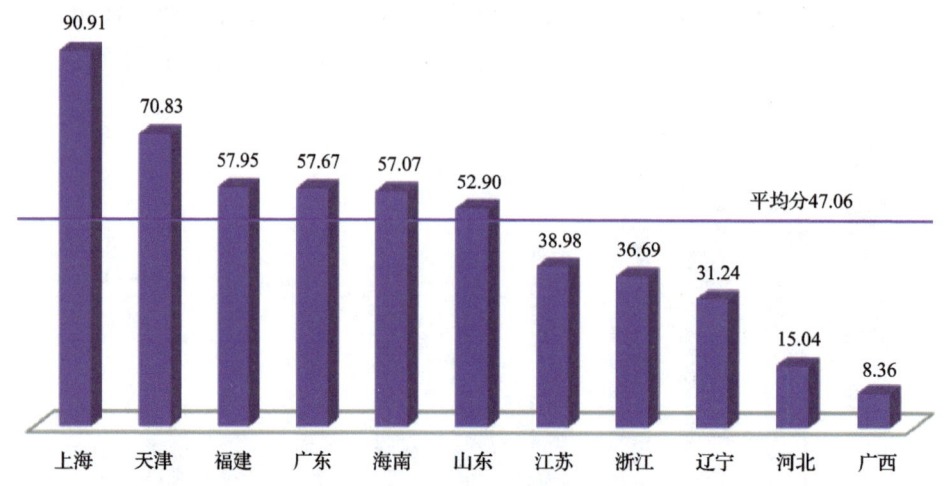

图 3-4　2016 年 11 个沿海省(自治区、直辖市)区域海洋创新绩效分指数得分及平均分

从海洋创新环境分指数来看，2016 年，得分超过平均分的沿海省(自治区、直辖市)有山东、福建、天津、上海、广西和广东(图 3-5)。其中，山东拥有良好的海洋创新人才环境和政府资金环境，其区域海洋创新环境分指数得分为 71.25，高于其他地区；福建的区域海洋创新环境分指数得分为 63.37，这得益于其优越的海洋设备和政府资金环境；天津得分为 59.03，这得益于其拥有较好的海洋创新资金环境和较高的人均海洋生产总值；上海得分为 56.14，这得益于其拥有很高的人均海洋生产总值；广西得分为 49.08，这得益于其优越的海洋创新资金环境；广东得分为 44.85，这得益于其拥有良好的政府资金环境。

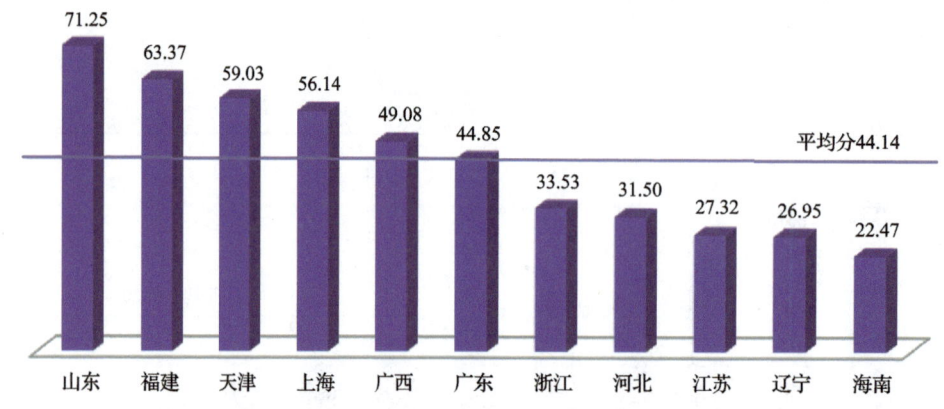

图 3-5　2016 年 11 个沿海省(自治区、直辖市)区域海洋创新环境分指数得分及平均分

二、区域海洋创新能力与经济发展水平强相关

由反映经济发展水平的"沿海地区人均生产总值"与"区域海洋创新指数"关系示意图(图 3-6)

可知，第一象限中的沿海地区人均生产总值较高，区域海洋创新指数高于全国平均水平，均为第一和第二梯次的地区；第四象限中的沿海地区人均生产总值相对较高，但区域海洋创新指数低于全国平均水平，除浙江外，均为第三梯次的地区；第三象限中河北、海南和广西人均生产总值相对较低、区域海洋创新指数低于全国平均水平，都是第四梯次的地区；没有一个地区处于人均生产总值较低，但区域海洋创新指数高于全国平均水平的第二象限。上述结果表明区域海洋创新能力与沿海区域经济发展水平之间具有强相关性。

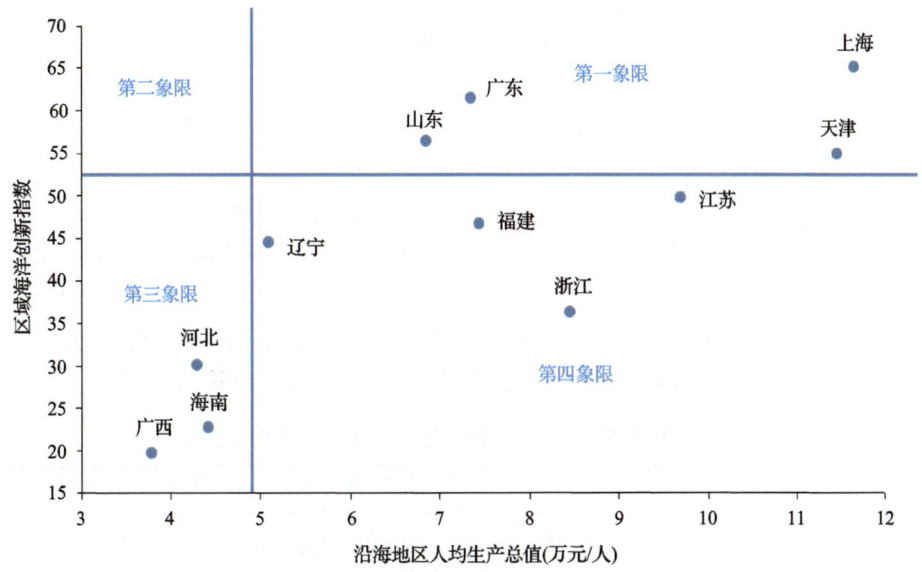

图 3-6　2016 年 11 个沿海省（自治区、直辖市）人均生产总值与区域海洋创新指数关系示意图

由反映海洋经济发展水平的"沿海地区人均海洋生产总值"与"区域海洋创新指数"关系示意图（图 3-7）可见，第一象限中的沿海地区人均海洋生产总值较高，区域海洋创新指数高于全国平

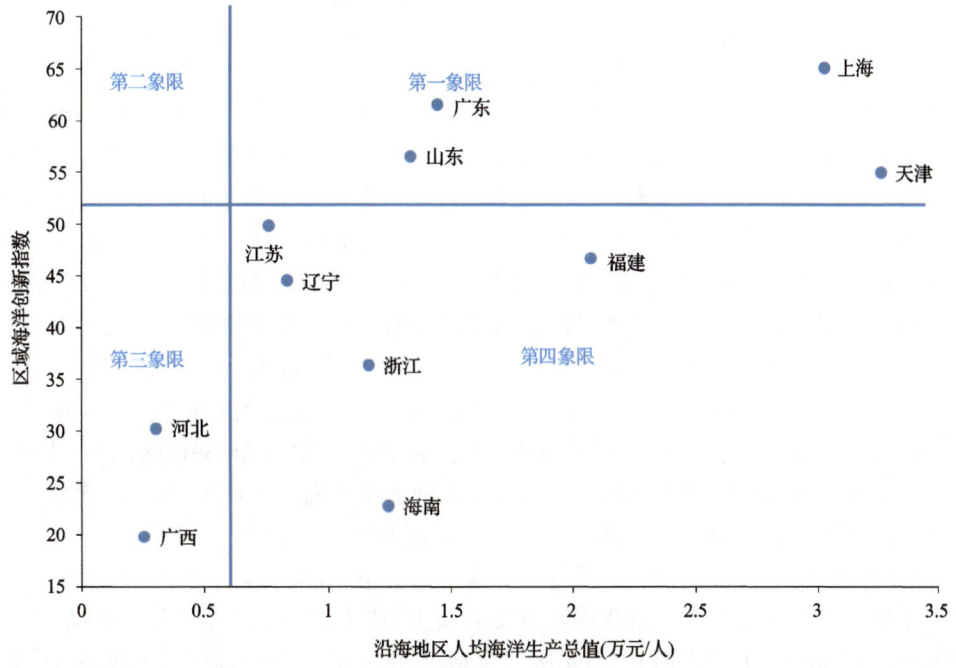

图 3-7　2016 年 11 个沿海省（自治区、直辖市）人均海洋生产总值与区域海洋创新指数关系示意图

均水平,也均是第一和第二梯次的地区;第四象限中沿海地区人均海洋生产总值相对较高,区域海洋创新指数接近或者低于全国平均水平,包含上述处于第三梯次所有地区及第四梯次的浙江和海南;第三象限中沿海地区人均海洋生产总值和区域海洋创新指数均低于全国平均水平,为第四梯次的河北和广西;没有一个地区处于沿海地区人均海洋生产总值低于全国平均水平,但区域海洋创新指数高于全国平均水平的第二象限。上述结果表明海洋创新活动与沿海区域海洋经济发展水平之间也具有强相关性。

第二节　从五大经济区看我国区域海洋创新发展

针对环渤海经济区、长江三角洲经济区、海峡西岸经济区、珠江三角洲经济区和环北部湾经济区五大经济区的具体分析如下。

环渤海经济区是指环绕着渤海全部及黄海的部分沿岸地区所组成的广大经济区域,是我国东部的"黄金海岸",具有相当完善的工业基础、丰富的自然资源、雄厚的科技力量和便捷的交通条件,也是我国中西部发展的战略地区,在全国经济发展格局中占有举足轻重的地位。2016 年,环渤海经济区的区域海洋创新指数为 46.50(表 3-2),略高于 11 个沿海省(自治区、直辖市)的平均水平,但区域海洋知识创造和海洋创新绩效在平均水平之下,海洋创新发展有提升的空间。

表 3-2　2016 年我国五大经济区区域海洋创新指数与分指数

经济区	综合指数	分指数			
	区域海洋创新指数 a	海洋创新资源 b_1	海洋知识创造 b_2	海洋创新绩效 b_3	海洋创新环境 b_4
环渤海经济区	46.50	53.43	42.90	42.50	47.18
长江三角洲经济区	50.41	60.81	46.29	55.52	39.00
海峡西岸经济区	46.69	34.02	31.42	57.95	63.37
珠江三角洲经济区	61.51	47.88	95.64	57.67	44.85
环北部湾经济区	21.22	8.56	7.84	32.72	35.78
平均值	45.27	40.94	44.82	49.27	46.04

长江三角洲经济区位于我国东部沿海、沿江地带交汇处,区位优势突出,经济实力雄厚。长江三角洲经济区以上海为核心,以技术型工业为主,技术力量雄厚、前景好、政府支持力度大、环境优越、教育发展好、人才资源充足,是我国最具发展活力的沿海地区。2016 年,长江三角洲经济区的区域海洋创新指数为 50.41,高于 11 个沿海省(自治区、直辖市)的平均水平,大量的海洋创新资源为长江三角洲经济区海洋科技与经济发展创造了良好的条件,海洋创新成果显著。

海峡西岸经济区以福建为主体包括周边地区,南北与珠江三角洲、长江三角洲两个经济区衔接,东与台湾、西与江西的广大内陆腹地贯通,是具备独特优势的地域经济综合体,具有带动全国经济走向世界的特点。2016 年,海峡西岸经济区的区域海洋创新指数为 46.69,略高于 11 个沿海省(自治区、直辖市)的平均水平,区域海洋创新环境与海洋创新绩效高于平均水平,有着较好的发展潜质,但海洋创新资源与海洋知识创造水平较低,海洋创新发展能力有待进一步提升。

珠江三角洲经济区主要指我国大陆南部的广东,与香港、澳门两大特别行政区接壤,科技力量与人才资源雄厚,海洋资源丰富,是我国经济发展最快的地区之一。珠江三角洲经济区的区域海洋创新指数为 61.51,高于 11 个沿海省(自治区、直辖市)的平均水平,且在五大经济区中位居首位,海洋创新资源密集，知识创造硕果累累，创新绩效成绩斐然。

环北部湾经济区地处华南经济圈、西南经济圈和东盟经济圈的结合部，是我国西部大开发地区中唯一的沿海区域，也是我国与东南亚国家联盟成员国既有海上通道又有陆地接壤的区域，区位优势明显，战略地位突出。环北部湾经济区的区域海洋创新指数为21.22，远低于11个沿海省(自治区、直辖市)的平均水平，在五大经济区中居末位，创新指数的4个分指数得分均比较低，与长江三角洲经济区及珠江三角洲经济区的差距较大。

第三节 从三大海洋经济圈看我国区域海洋创新发展

2016年，东部海洋经济圈的区域海洋创新指数为50.41，居三大海洋经济圈之首(表3-3，图3-8)。4个分指数中得分较高的是海洋创新资源分指数和海洋创新绩效分指数，分别为60.81和55.52，这两个分指数对该经济圈的区域海洋创新指数有较大的正贡献，充分说明该区域优势突出、经济实力雄厚，其优质的海洋创新资源为区域海洋科技与经济发展创造了良好的条件；得分较低的是海洋知识创造和海洋创新环境分指数，分别为46.29和39.00，这两个分指数对区域海洋创新指数有负效应(图3-9)。

北部海洋经济圈的区域海洋创新指数为46.50，得分在三大海洋经济圈居中。4个分指数中，海洋创新资源和海洋创新环境对区域海洋创新指数有正贡献作用，得分分别为53.43和47.18；海洋知识创造和海洋创新绩效的得分比较低，分别为42.90和42.50。北部海洋经济圈的区域海洋创新指数得分较低的原因主要在于海洋创新绩效相对较弱，海洋创新发展有待进一步提高。

表3-3 2016年我国三大海洋经济圈区域海洋创新指数与分指数

经济圈	综合指数	分指数			
	区域海洋创新指数 a	海洋创新资源 b_1	海洋知识创造 b_2	海洋创新绩效 b_3	海洋创新环境 b_4
北部海洋经济圈	46.50	53.43	42.90	42.50	47.18
东部海洋经济圈	50.41	60.81	46.29	55.52	39.00
南部海洋经济圈	37.66	24.75	35.68	45.26	44.94

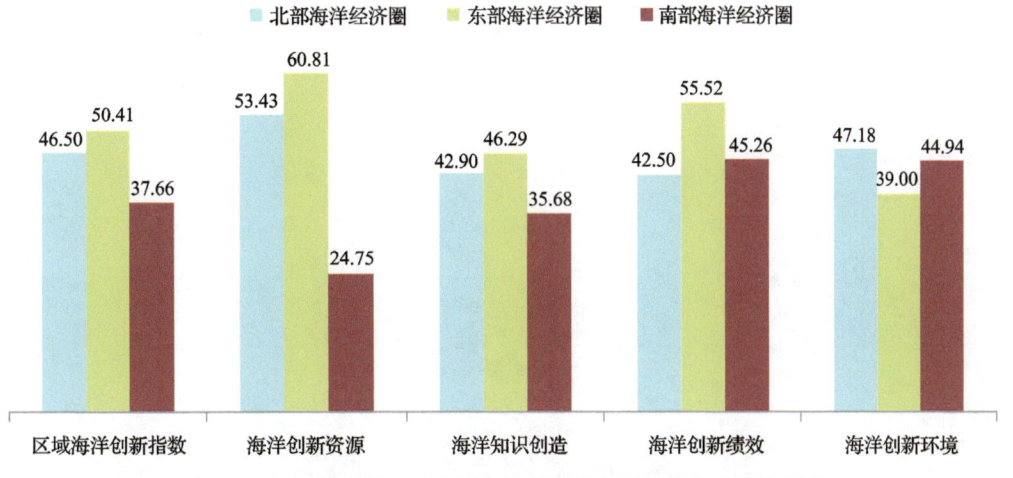

图3-8 2016年我国三大海洋经济圈海洋创新指数与分指数得分

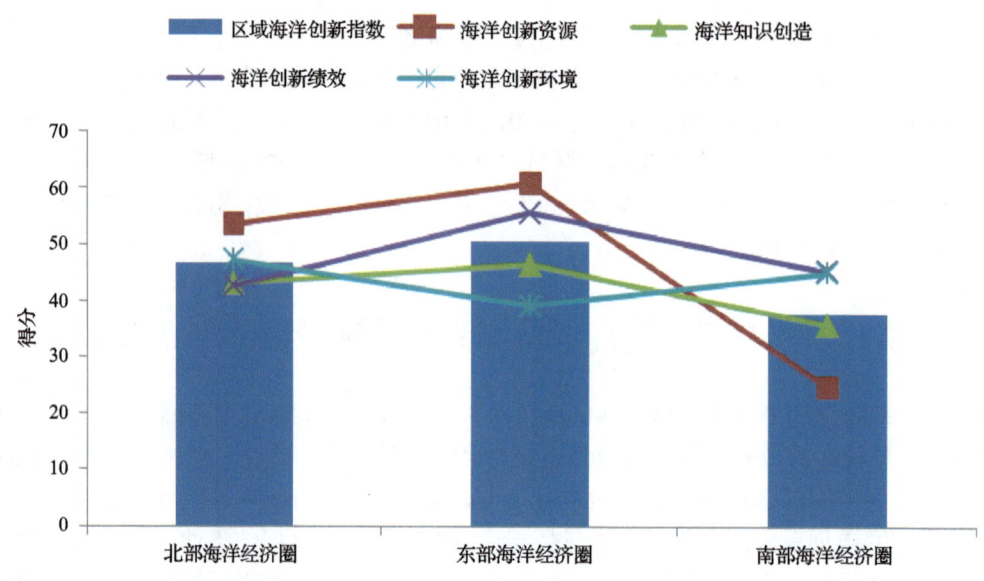

图 3-9 2016 年我国三大海洋经济圈区域海洋创新指数与分指数关系

南部海洋经济圈的区域海洋创新指数为 37.66，在三大海洋经济圈中最低。4 个分指数得分差异较大，其中，海洋创新环境和海洋创新绩效两个分指数得分较高，分别为 44.94 和 45.26；海洋知识创造和海洋创新资源分指数得分较低，分别为 35.68 和 24.75，是造成区域海洋创新指数较低的主要因素。南部海洋经济圈在三大海洋经济圈中得分最低，提升空间较大。在以后的海洋创新发展过程中，需要进一步发挥珠江口及两翼的创新总体优势，带动福建、北部湾和海南岛沿岸发挥区位优势共同发展，使海洋创新驱动经济发展的模式辐射至整个南部海洋经济圈。

第四章 我国海洋创新能力的进步与展望

习近平总书记强调"要发展海洋科学技术,着力推动海洋科技向创新引领型转变。""要依靠科技进步和创新,努力突破制约海洋经济发展和海洋生态保护的科技瓶颈。要搞好海洋科技创新总体规划"。创新是引领经济增长最为重要的引擎,海洋创新更是指导海洋事业不断突破、实现海洋经济稳步健康发展的重要支撑。

国家海洋创新能力与海洋经济发展相辅相成。我国海洋创新能力的提高,与海洋经济发展相互关联。2012~2016年国家海洋创新指数、海洋生产总值和国内生产总值的增长率接近,国家海洋创新能力基本与海洋经济发展水平保持一致,海洋创新对经济的贡献能力也同步提升。

"十三五"开局之年,海洋科学和技术发展的部分指标已接近预期规划目标,发展态势良好。2016年,海洋生产总值占国内生产总值比重达到9.51%,海洋科技进步贡献率达到65.9%,超过预期规划目标;科技成果转化率达到50.0%,与规划目标的55.0%有一定差距,科技创新成果转化能力仍有较大提升空间。

第一节　国家海洋创新能力与海洋经济发展相辅相成

国家海洋创新能力与海洋经济发展相辅相成，海洋经济为海洋科技研发提供充足的资金保障，从而提高海洋资源利用效率；海洋科技的进步和创新能力的提高，又促进海洋经济和国民经济的增长。2004～2016 年国家海洋创新指数、海洋生产总值和国内生产总值总体均呈现增长趋势（图 4-1），年均增长率分别为 7.58%、13.98% 和 13.63%（表 4-1）。国家海洋创新能力与海洋经济发展基本趋势保持一致，但国家海洋创新指数增长率不及海洋生产总值和国内生产总值，这说明国家海洋创新对经济增长的贡献还存在较大的发展空间。

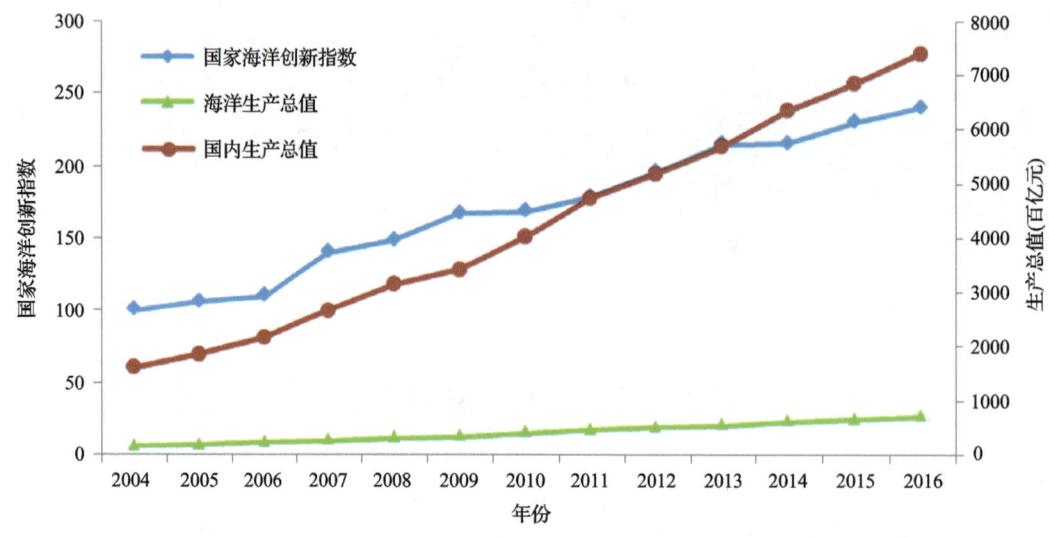

图 4-1　2004～2016 年国家海洋创新指数、海洋生产总值与国内生产总值

表 4-1　国家海洋创新指数、海洋生产总值与国内生产总值增长率（%）

年份	国家海洋创新指数增长率	海洋生产总值增长率	国内生产总值增长率
2004	—	—	—
2005	5.61	20.42	15.67
2006	3.87	22.30	16.97
2007	27.52	18.65	22.88
2008	5.93	16.00	18.15
2009	12.59	8.61	8.55
2010	0.94	22.60	17.78
2011	5.93	14.97	17.83
2012	9.82	10.00	9.69
2013	9.91	8.53	9.62
2014	0.28	11.76	11.83
2015	6.79	6.54	7.76
2016	4.27	9.03	8.12
年均增长率	7.58	13.98	13.63

第二节 国家海洋"十三五"相关规划重要指标进展

《全国科技兴海规划(2016—2020年)》和《全国海洋经济发展"十三五"规划》等对"十三五"期间的海洋创新发展提出了明确要求,旨在引领"十三五"我国海洋创新发展。在"十三五"开端,对这些目标的实际情况进行数据分析,为"十三五"管理部门及时掌握国家海洋创新能力情况和发展趋势提供依据。

2016年,海洋生产总值占国内生产总值比重达到9.51%,海洋科技进步贡献率达到65.9%;海洋科技成果转化率达到50.0%,与规划目标的55.0%有一定差距,科技创新成果转化能力仍有较大提升空间。2016年是"十三五"开局之年,部分指标达到规划目标并呈现上升趋势(表4-2),发展态势良好。

表4-2 国家海洋"十三五"相关规划重要指标情况(%)

主要指标	2015年	"十三五"目标	实际测算值
海洋生产总值占国内生产总值比重	9.4	9.5	9.51(2016年)
海洋科技进步贡献率	>60	>60	65.9(2006~2016年)
海洋科技成果转化率	>50	>55	50.0(2000~2016年)

展望未来,我国应进一步加大海洋创新资源投入力度,同时注重海洋创新的效率问题,发挥海洋创新的支撑引领作用,转变海洋经济发展方式,推动海洋经济转型升级,依靠海洋科技突破经济社会发展中的能源、资源与环境约束,让海洋创新成为驱动海洋经济发展与转型升级的核心力量,为海洋强国建设提供充足的知识储备和坚实的技术基础。

第五章 我国海洋科研机构的空间分布特征与演化趋势

海洋科研机构是国家海洋创新发展的主要动力和国家海洋科研能力建设的重要部分。在建设海洋强国战略背景下，科学谋划海洋科研机构空间布局、合理配置海洋科技创新资源具有重大意义。

为揭示中国海洋科研机构布局和海洋科研力量的空间演化规律，本章基于海洋科技统计中海洋科研机构单点数据，采用标准差椭圆这一空间统计方法，选取科技统计中的从业人员、科技活动收入中政府资金和科技论文3个指标分别反映海洋科研机构的人力投入、经费投入和产出情况，多重角度刻画海洋科研机构的地理位置、人力投入、经费投入、产出等要素的时空变化过程，以空间可视化的方式揭示我国海洋科研机构格局的整体特征与动态演化过程，以期为制定海洋科技创新发展政策、海洋科研机构布局战略等提供决策依据。

第一节 研究方法

标准差椭圆(standard deviational ellipse, SDE)是空间统计方法中能够从多重角度反映要素空间分布整体性特征的方法,已成为ArcGIS空间统计模块的常规统计工具。

SDE由美国南加利福尼亚大学社会学教授Lefever于1926年提出,用来度量一组数据的方向和分布,揭示要素的空间分布特征,因其直观性与有效性而得到广泛应用。SDE方法生成的结果为一个椭圆,从其生成算法来看,首先用平均中心来确定椭圆的圆心,然后由平均中心作为起点对X坐标和Y坐标的标准差进行计算,从而定义椭圆的轴线,同时确定椭圆的方向,正北方向为$0°$,顺时针旋转。此外,可以根据要素的位置点或受与要素关联的某个属性值影响的位置点来计算标准差椭圆。需要说明的是,ArcGIS提供了"椭圆大小"这一参数,一个、两个、三个标准差范围可分别将约占总数68%、95%、99%的输入要素包含在椭圆内。本研究选取一个标准差范围计算海洋科研机构的地理空间标准差椭圆(以下称geo椭圆),且选取从业人员、科技活动收入中政府资金和科技论文作为属性值计算海洋科研机构的加权标准差椭圆(以下分别称pe椭圆、fi椭圆和ot椭圆)。

SDE方法通过椭圆的空间分布范围和中心、长轴、短轴、方位角等基本参数定量描述海洋科研机构的要素空间分布特征。椭圆空间分布范围的含义与其是否设定属性值有关。例如,geo椭圆的空间分布范围直接表示海洋科研机构在地理位置上分布的主体区域,pe椭圆是以从业人员为权重的海洋科研机构输出的椭圆,其空间分布范围与geo椭圆的相对位置可反映海洋科研机构从业人员的空间分布情况。椭圆中心表示要素空间分布的平均中心;长轴的方向是要素空间分布的主趋势方向,其长度反映要素在主趋势方向上的离散程度,短轴反映要素空间分布的范围,长短轴比值越大,数据呈现的向心力越明显,反之,数据的离散程度越大;方位角是从正北方向顺时针旋转到椭圆长轴的角度,表征要素空间分布的方向。

第二节 空间分布特征与演化趋势

一、总体概况

2001~2015年中国海洋科研机构的geo椭圆、pe椭圆、fi椭圆和ot椭圆如图5-1所示。为辅助理解中国海洋科研机构在地理位置、人力投入、经费投入、产出上的空间分布特征和动态演化,选取2001年、2008年和2015年的标准差椭圆制图(图5-2,图5-3)。

从地理位置来看(图5-2),中国海洋科研机构的标准差椭圆的主趋势方向均为北偏东、南偏西,呈狭长分布。具体来说,2001年,geo椭圆空间分布从北到南覆盖渤海湾海域、山东、江苏、上海、安徽、浙江、江西、福建,该区域为2001年海洋科研机构在地理位置上分布的主体区域。相比于geo椭圆,pe椭圆和fi椭圆方位角明显偏小,且空间分布往北集中,这是由于位于北京的机构在人力投入和经费投入方面有明显优势。ot椭圆方位角变化不大,其空间分布在geo椭圆内部往西集中,说明北方的海洋科研机构在产出上不占优势,且位于中部(如湖北)的海洋科研机构虽然数量不多,但产出可观。

2008年,中国广西、海南等沿海地区和湖北、陕西、甘肃等中西部地区建设起一批海洋科研机构,geo椭圆相对2001年向西南方向扩展。相比于geo椭圆,pe椭圆、fi椭圆和ot椭圆空间分布往北集中,且方位角变小,其中,*ot* 椭圆程度最大,且整体更偏向于西。这是因为国家大幅增加了北京、山东及中西部地区海洋科研机构的人力、经费投入,北京、天津、山东、广东等地区海洋科研机构的产出数据明显领先于其他地区。

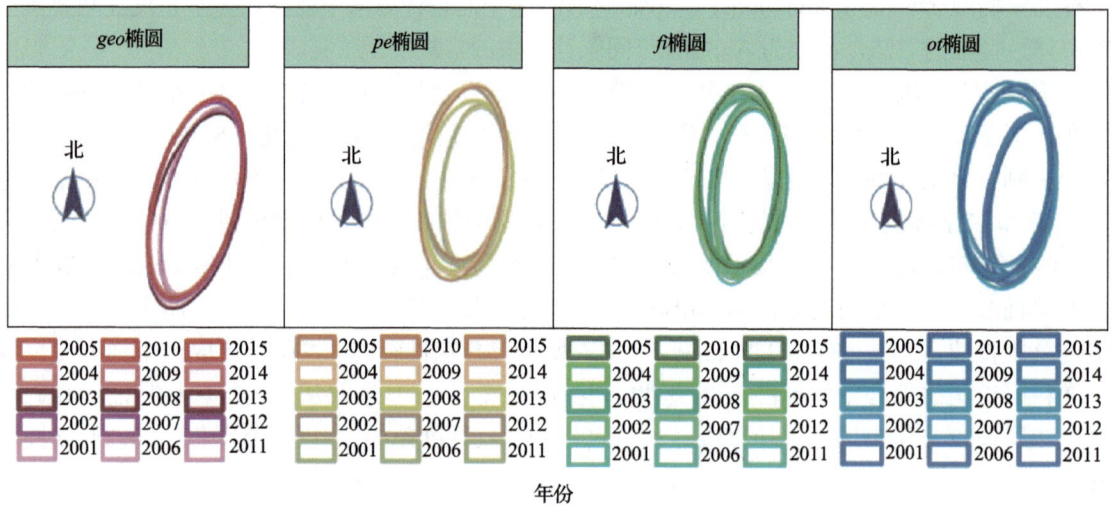

图 5-1　2001～2015 年中国海洋科研机构标准差椭圆

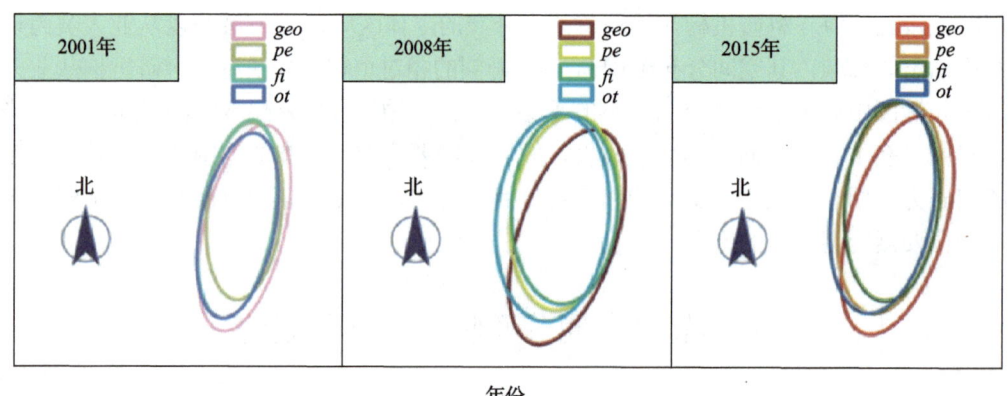

图 5-2　2001 年、2008 年、2015 年中国海洋科研机构标准差椭圆(分年份)

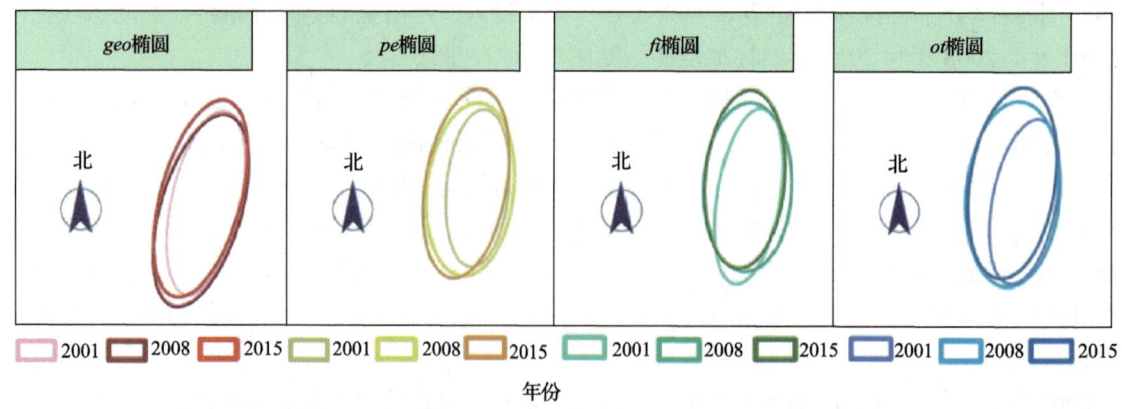

图 5-3　2001 年、2008 年、2015 年中国海洋科研机构标准差椭圆(分权重)

相比于 2008 年，2015 年 geo 椭圆向北移动，中国海洋科研机构在全国范围内有了新的空间格局，其中，黑龙江、辽宁、北京等北方区域在数量上涨幅较大。从 4 个椭圆的相对位置来看，其关系与 2008 年大致相同，fi 椭圆往北集中趋势更为明显，ot 椭圆也往北移动，趋向于以山东为中心区域，这说明该时段北京、天津、山东、浙江等地的经费投入和论文产出依然占据优势。

从动态变化过程来看（图 5-3），geo 椭圆的空间分布先往西南方向扩展，再向北移动。pe 椭圆先有整体的扩展，往西和往南最为明显，然后在继续扩展中有偏向于北的趋势，表明中国在全国范围内不断加强海洋科研人才建设，有效促进了广西等沿海地区和中西部地区的人才引进。fi 椭圆的扩展以往北为主，往西为辅，说明中国海洋科研经费投入仍集中在北京、天津、山东、浙江等老牌海洋强省（直辖市）强市，同时，中西部的海洋科研机构也得到了政府的大力支持。ot 椭圆的整体扩展趋势最为明显，与 fi 椭圆一样以往北扩展为主，往西偏移为辅，不同的是，ot 椭圆的扩展基本完全基于自身，而 fi 椭圆则是扩展的同时往北移动，反映了海洋科研产出在全国范围内的大爆发，这说明中国海洋科研活动程度不断提高，海洋科技创新水平持续提升。

二、平均中心变化趋势

标准差椭圆的中心表示要素空间分布的平均中心。从中心点的轨迹（图 5-4）来看，2001~2015 年，中国海洋科研机构标准差椭圆有明显变化。为便于总结海洋科研机构建设的演变规律，暂不考虑经纬度关系，制图 5-5；为便于明晰海洋科研机构的地理位置与人力投入、经费投入和产出的关系，制图 5-6。

从图 5-4 可以看到，geo 椭圆中心点的轨迹可分为四段：①2001~2006 年，该时间段内中国海洋科研机构大多集中在沿海地区，且变化不大；②2007~2009 年，该时间段是中国海洋科研机构建设的全面爆发期，黑龙江、甘肃、陕西、湖北、海南等地均建设有海洋科研机构，且沿海地区海洋科研机构数量也在增加，geo 椭圆中心点变动较大，总体向西移动；③2010~2012 年，geo 椭圆中心点继续向西移动，处于稳定发展状态；④2013~2015 年，北京、天津、山东等地海洋科研机构数量增多，geo 椭圆中心向北移动且趋于稳定。

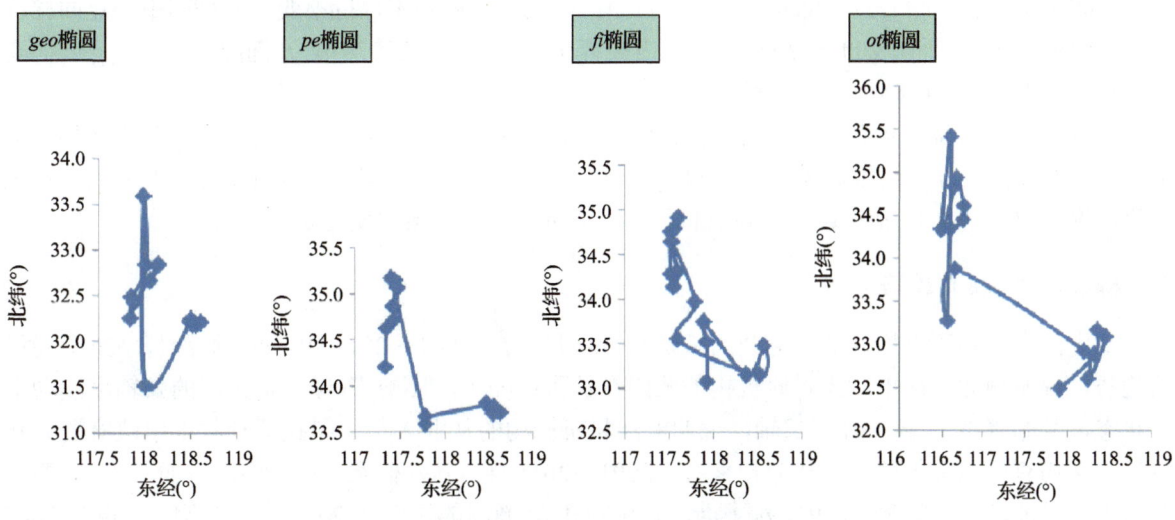

图 5-4　2001~2015 年中国海洋科研机构标准差椭圆中心点轨迹图

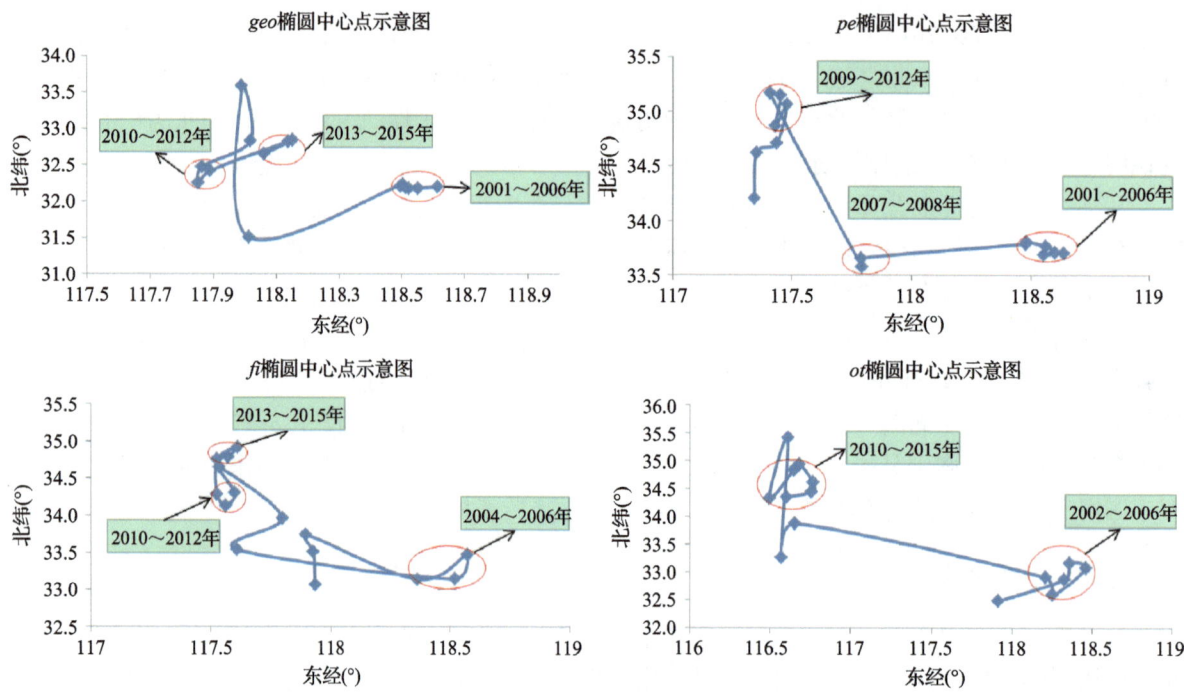

图 5-5 2001~2015 年中国海洋科研机构标准差椭圆中心点示意图(分权重)

pe 椭圆、fi 椭圆、ot 椭圆中心点的轨迹需要与 geo 椭圆中心点轨迹结合探讨。对 pe 椭圆来说，2001~2006 年，pe 椭圆中心点变动不大，与 geo 椭圆中心点的相对位置也基本保持不变，位于 geo 椭圆中心点的正北方，说明北京、天津、山东等地区的海洋科研机构从业人员有明显优势；2007 年，pe 椭圆中心点向西移动；2009 年，pe 椭圆中心点继续向西北方向移动；2013~2015 年，pe 椭圆中心点持续向南移动，且 2007~2015 年，pe 椭圆中心点始终位于 geo 椭圆中心点的西北方向，说明中西部地区的海洋科研机构人才建设环境良好，且 2013 年以后，南方海洋科研机构逐步增加了从业人员。对 fi 椭圆来说，fi 椭圆中心点始终位于 geo 椭圆的西北方向，2001~2008 年，两者距离越来越小，2009 年，fi 椭圆中心点与 geo 椭圆中心点距离拉大，此后基本保持不变。ot 椭圆中心点同样位于 geo 椭圆的西北方向，2001~2006 年，两者距离呈波动状态，2007 年，fi 椭圆中心点与 geo 椭圆中心点距离大幅拉大，此后有缓慢变小趋势。

pe 椭圆、fi 椭圆、ot 椭圆中心点与 geo 椭圆中心点的空间关系可以为确定海洋科研机构科研力量的均衡性提供参考。整体来看，pe 椭圆、fi 椭圆、ot 椭圆中心点与 geo 椭圆中心点的距离有扩大趋势，说明 2001~2015 年中国海洋科研机构科研力量呈现非均衡化变化趋势。

三、椭圆长短轴变化趋势

从中国海洋科研机构标准差椭圆的长轴来看(图 5-7)，2001~2015 年，geo 椭圆长轴呈缓慢增长趋势，说明南北方向上海洋科研机构的范围在扩张；pe 椭圆、fi 椭圆、ot 椭圆的长轴呈波动上升状态，且始终小于 geo 椭圆长轴，说明海洋科研机构的从业人员、科技活动收入中政府资金和科技论文均有明显聚集现象。从短轴来看，2001~2015 年，geo 椭圆、pe 椭圆、fi 椭圆和 ot 椭圆的短轴均呈波动增长趋势，其中，pe 椭圆、fi 椭圆和 ot 椭圆的短轴在 2007 年大幅增长，说明 2007 年在黑龙江、甘肃、海南等地区建设的海洋科研机构在人力投入、经费投入和产出方面都有明显优势。

第五章 我国海洋科研机构的空间分布特征与演化趋势

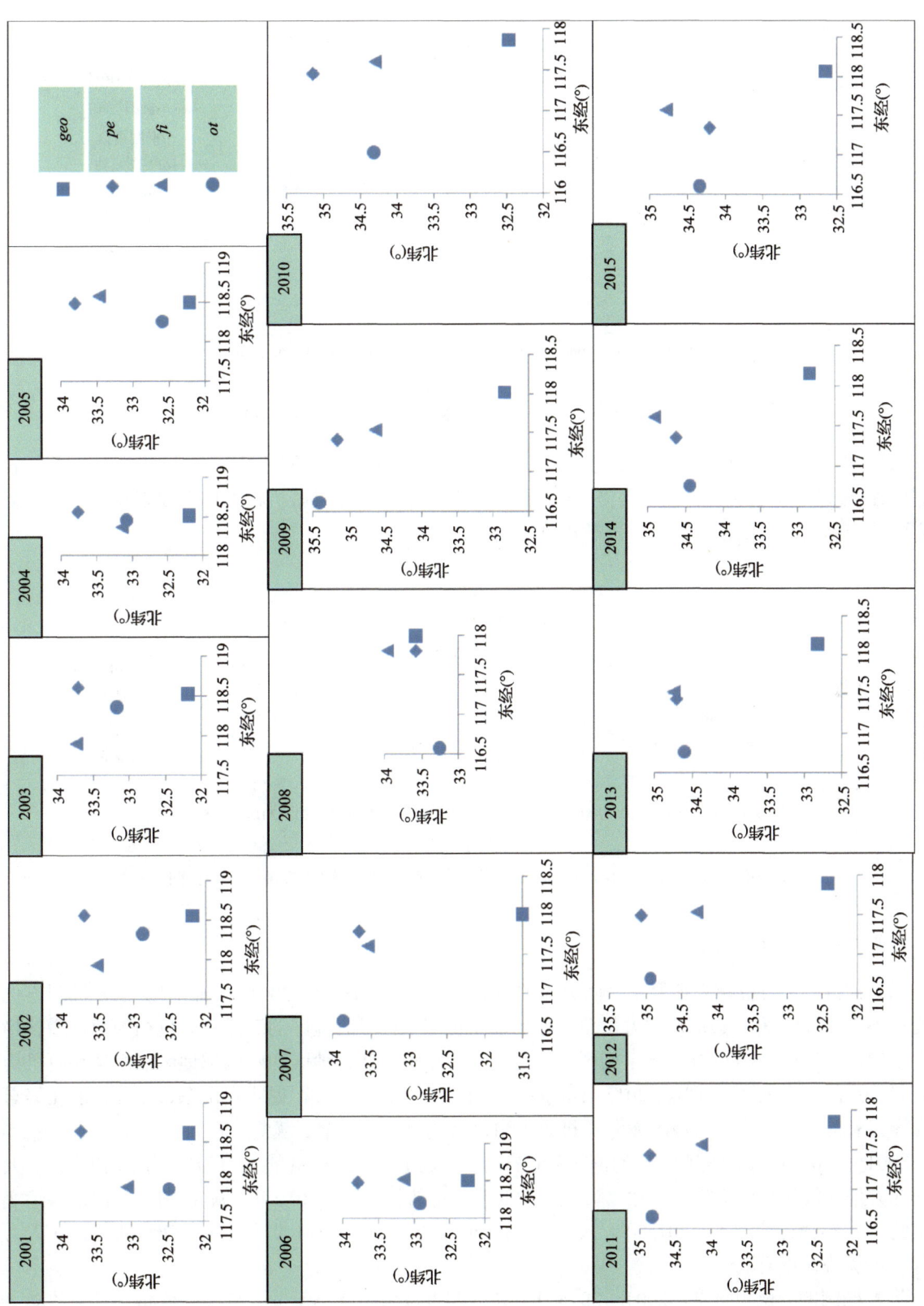

图 5-6 2001～2015 年中国海洋科研机构标准差椭圆中心点示意图（分年份）

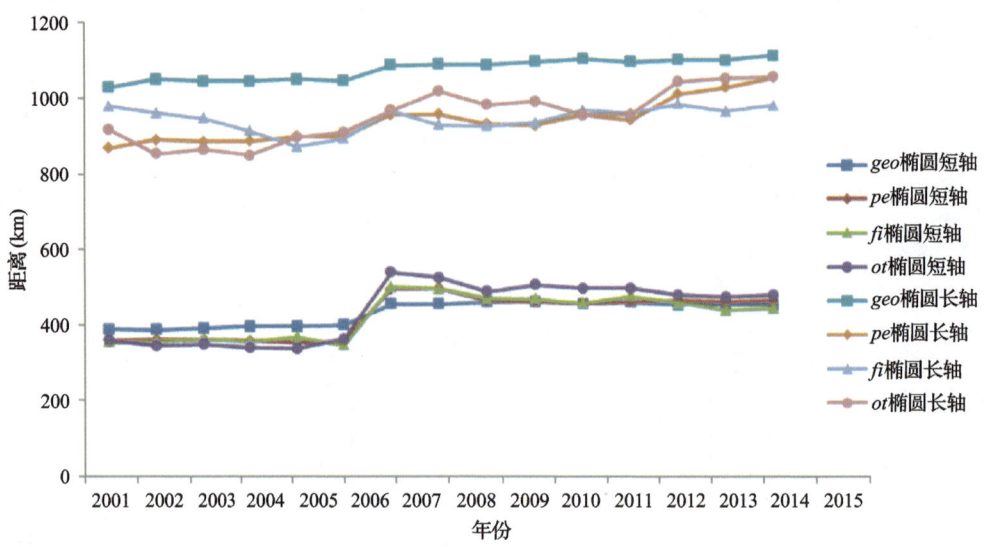

图 5-7　2001~2015 年中国海洋科研机构标准差椭圆长短轴趋势图

从短轴与长轴的比值来看（图 5-8），geo 椭圆短轴与长轴比值缓慢增长，pe 椭圆、fi 椭圆和 ot 椭圆则在 2007 年涨势突出，此后缓慢下降。总体来说，中国海洋科研机构在地理要素和科研力量要素方面的空间展布范围都在扩张，离散程度增大。

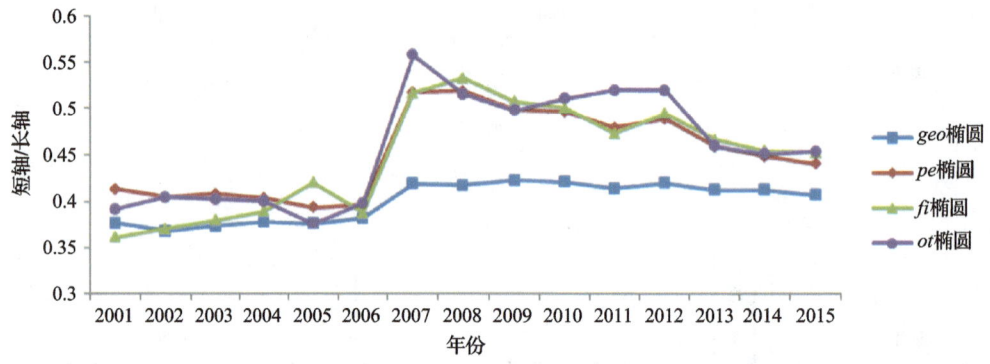

图 5-8　2001~2015 年中国海洋科研机构标准差椭圆短轴与长轴比值趋势图

四、椭圆方位角变化趋势

标准差椭圆的方位角表征要素空间分布的方向。如图 5-9 所示，2001~2006 年，geo 椭圆方位角变动不大；2007 年，geo 椭圆方位角快速增长，说明东南沿海地区的海洋科研机构的数量增长趋势大于中西部地区；2009 年，geo 椭圆方位角变小，反映出北方地区和中西部地区的海洋科研机构增长趋势反超东南地区；2010~2015 年，geo 椭圆方位角缓慢增大，说明在山东、江苏、浙江等老牌海洋强省的聚集作用下，该区域海洋科研机构在数量上保持了增长优势。

2001~2008 年，pe 椭圆的方位角保持在 7°左右，远小于 geo 椭圆的 12°~17°，显示出北京、天津等北方地区海洋科研机构在从业人员数量上的明显优势；2009~2010 年，pe 椭圆的方位角持续变小至约 4°，继而持续增大，2015 年，pe 椭圆的方位角约为 10°，说明北方地区海洋科研机构在人力投入上的优势在逐渐减弱。

从 fi 椭圆的方位角来看，2001~2015 年，fi 椭圆方位角始终小于 geo 椭圆方位角，且呈波动下降趋势。一方面，北京、天津等北方地区海洋科研机构在政府经费上远高于其他地区；另一方面，中西部地区的海洋科研机构在经费上的拉动作用十分明显。

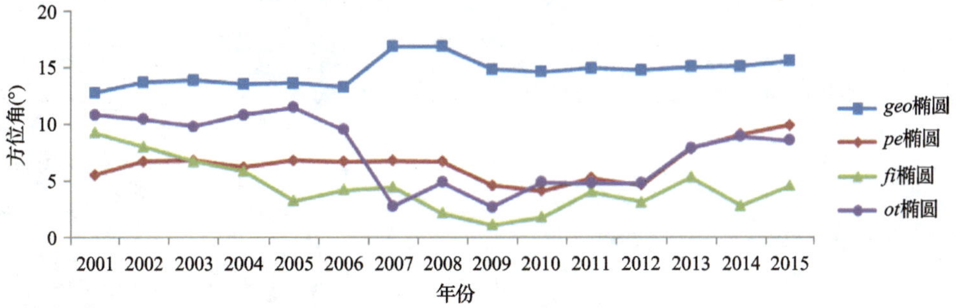

图 5-9 2001~2015 年中国海洋科研机构标准差椭圆长轴绕圆心北偏东方位角趋势图

2001~2015 年，ot 椭圆的方位角大体呈 "U" 形趋势。2007 年，ot 椭圆的方位角大大减小，说明中西部地区和广东、广西等地区的海洋科研机构的科技论文发表数量飞速增长，反映出这些地区海洋科技原始创新能力的提高。此后，ot 椭圆的方位角波动上升，至 2013 年明显大幅提升后恢复平稳增长，说明东南沿海地区科技论文发表量的增长趋势实现反超，全国范围内海洋科研机构产出增长趋势趋同。

第三节 主要研究结论

进入 21 世纪以来，海洋科技创新成为海洋经济发展的根本动力，在国家、区域竞争中占据主要地位。作为海洋科技创新的主力军之一，海洋科研机构的地理要素和科研力量要素的空间布局对落实建设海洋强国意义重大。本章揭示了 2001~2015 年海洋科研机构的空间分布特征与动态演化过程，可得出以下结论。

（1）从地理位置来看，中国海洋科研机构数量持续增长，空间展布范围不断扩张。具体来说，中国海洋科研机构在地理位置上大多集中在沿海地区；从 2007 年开始，黑龙江等北方地区，湖北、陕西、甘肃等中西部地区和广西、海南等沿海地区建设起一批海洋科研机构，但其增长趋势小于东南沿海地区的海洋科研机构；2009 年，北方地区和中西部地区的海洋科研机构增长趋势反超东南沿海地区；2010 年以后，这一趋势减缓，山东、江苏、浙江等地恢复增长优势。

（2）从人力投入来看，中国持续加强海洋科研人才建设。具体来说，北京、天津、山东等地区的海洋科研机构从业人员有明显优势；自 2007 年开始，中西部地区的海洋科研机构人才建设环境良好；2013 年以后，北方地区海洋科研机构在人力投入上的优势逐渐减弱，南方海洋科研机构的从业人员数量涨势突出。

（3）从经费投入来看，中国海洋科研经费投入集中在北京、天津、山东、浙江等老牌海洋强省（直辖市）强市。同时，中西部地区的海洋科研机构在经费上的拉动作用也十分明显。

（4）从科研产出来看，中国海洋科研机构的科研产出正在实现全国范围内的大爆发，海洋科研活动程度不断提高。具体来说，2007 年以前，北方的海洋科研机构在产出增长率上不占优势，而位于中部的海洋科研机构产出成果突出；2007 年，中西部地区和广东、广西等地区的海洋科研机构的科技论文发表数量飞速增长，北京、天津、辽宁等地区科研产出同样可观；自 2009 年开始，东南沿海地区科技论文发表量的增长趋势实现反超，海洋科研机构产出在全国范围内增长趋势明显。

（5）从四者空间关系来看，中国海洋科研机构的地理位置与人力投入、经费投入和科研产出布局呈现非均衡化趋势。2001~2015 年，中国海洋科研机构的人力投入、经费投入和科研产出与地理位置的布局有明显出入，且其中心点的距离有扩大趋势，呈非均衡化发展态势。

第六章 全球海洋创新能力分析

全球海洋领域 SCI 论文总量保持稳定增长态势。2016 年论文发表数量是 2001 年的 1.68 倍,年均增长率为 3.54%。

2001~2016 年全球海洋领域 SCI 发文数量前 15 位的国家分别为美国、中国、英国、澳大利亚、法国、德国、加拿大、日本、西班牙、俄罗斯、意大利、挪威、荷兰、印度和韩国。论文总被引频次最高的为美国,其次为英国、德国、法国、澳大利亚、加拿大,我国尽管论文总量排名第二,但是论文总被引频次却排名第七位。

基于 2001~2016 年全球海洋科技领域前 20 的机构发表 SCI 论文数量及年度变化分析,主要发文机构中 8 个机构为美国所属,3 个主要机构为中国所属,分别为中国科学院、中国海洋大学和原国家海洋局。

海洋领域 SCI 研究论文涉及学科众多且存在交叉,主要是海洋生物、海洋工程、海洋地球化学、渔业、环境、地质及采矿等相关领域。

海洋领域 EI 论文呈现快速增长趋势。中美两国海洋领域 EI 论文发表数量占全球 40%左右,年度论文增长幅度远高于其他国家。自 2011 年以来,中国 EI 论文产出量超过美国,位居全球首位。

中国海洋专利申请数量位居第一,遥遥领先于其他国家和地区,占世界比例稳步提高,从 2001 年的 5.4%提高至 2016 年 84.4%。世界海洋专利申请前 15 位主要机构中,有 5 家机构为中国所属,分别是中国海洋石油集团有限公司、浙江海洋大学、中国海洋大学、浙江大学和大连海洋大学。

第一节 全球海洋创新成果总量与态势分析

一、SCI 论文保持稳定增长态势

2001~2016 年，全球海洋创新 SCI 论文总量持续增长，2016 年论文发表数量是 2001 年的 1.68 倍，年均增长率为 3.54%。如图 6-1 所示，2001~2016 年论文数量呈现明显的三阶段变化，2006 年和 2012 年为转折点。

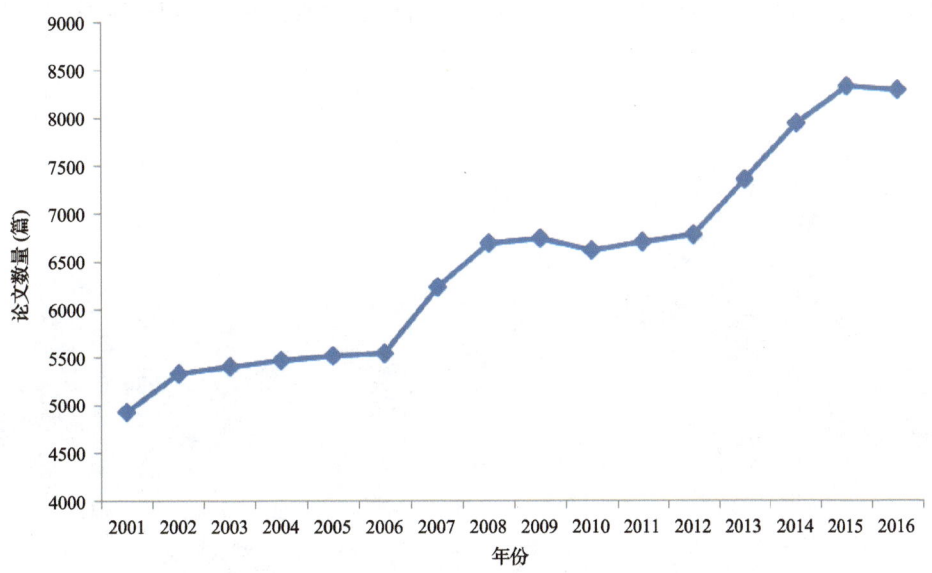

图 6-1 2001~2016 年全球海洋科技 SCI 论文发表数量年度变化

从 2001~2016 年全球海洋领域前 20 机构发表 SCI 论文数量（图 6-2）来看，发表论文数量最多的

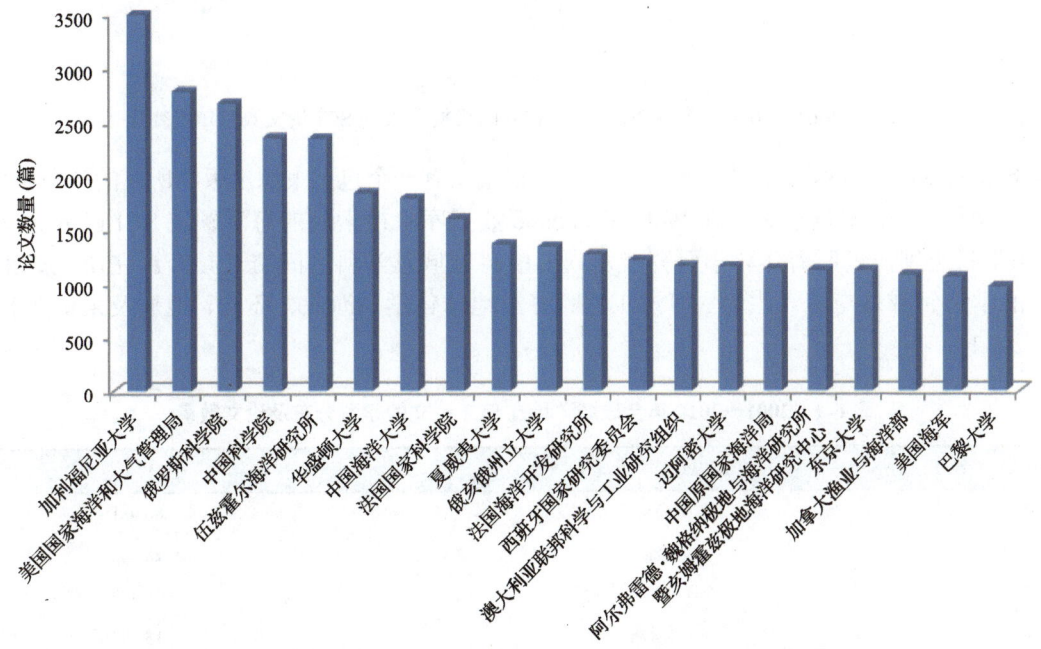

图 6-2 2001~2016 年全球前 20 机构发表海洋科技 SCI 论文数量

机构为美国的加利福尼亚大学，其次为美国国家海洋和大气管理局、俄罗斯科学院、中国科学院、伍兹霍尔海洋研究所、华盛顿大学、中国海洋大学、法国国家科学院、夏威夷大学和俄亥俄州立大学等。其中在前 20 的主要发文机构中 8 个机构为美国所属；3 个机构为中国所属，分别为中国科学院、中国海洋大学和原国家海洋局；3 个机构为法国所属；加拿大、德国、西班牙、日本、澳大利亚、俄罗斯所属机构均为 1 个。

2001~2016 年，全球海洋领域前 20 的机构发表的 SCI 论文数量年度变化情况如图 6-3 所示，中国机构在最近 3 年的发文量占主要优势。2016 年，夏威夷大学、西班牙国家研究委员会和阿尔弗雷德·魏格纳极地与海洋研究所暨亥姆霍兹极地海洋研究中心（AWI）发文量相对较少。

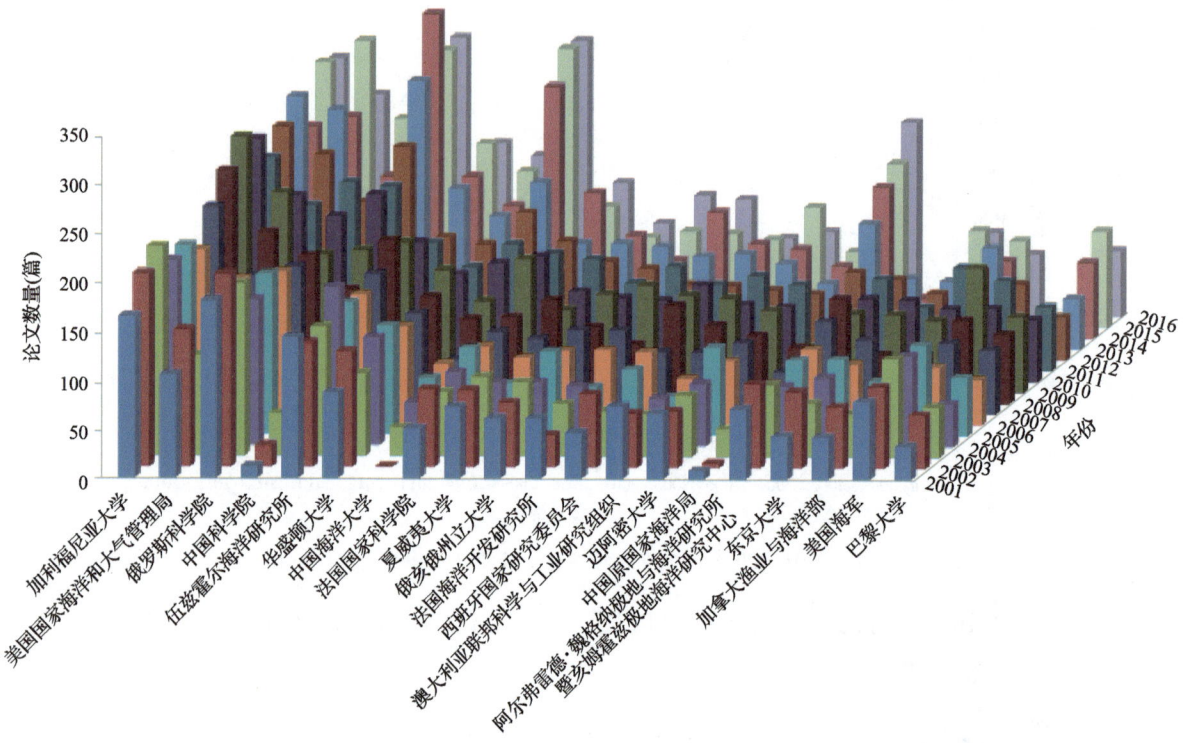

图 6-3　2001~2016 年全球海洋领域前 20 的机构发表的 SCI 论文数量年度变化

Web of Science（WOS）数据库中收录的每一条记录均有一个包含来源出版物所属的学科类别，覆盖 252 个学科类别。根据检索式在 Web of Science 数据库中检索到的海洋领域 SCI 研究论文共涉及 23 种学科类别，表明海洋科学研究涉及众多学科领域且学科之间交叉频繁。在研究成果中涉及较多的学科领域是与海洋生物、海洋工程、海洋地球化学、渔业、环境、地质及采矿等相关的领域（表 6-1）。

表 6-1　2001~2016 年全球海洋科技 SCI 论文的学科分布及论文数量

序号	WOS 学科分类	论文数量（篇）
1	海洋学	86 931
2	海洋工程	28 562
3	海洋与淡水生物学	28 493
4	土木工程	13 992
5	气象学和大气科学	9 409

续表

序号	WOS 学科分类	论文数量(篇)
6	生态学	8 916
7	地学交叉学科	8 284
8	湖沼学	7 174
9	渔业学	6 958
10	水资源学	4 337
11	机械工程	2 485
12	化学交叉学科	1 535
13	地球化学与地球物理学	1 344
14	古生物学	1 280
15	电子与电气工程	1 153
16	工程交叉学科	858
17	环境科学	541
18	力学	541
19	地质工程	429
20	采矿与选矿	429
21	遥感	422
22	动物学	199
23	能源和燃料	92

二、EI 论文快速增长

2001～2016 年全球海洋领域 EI 论文数量如图 6-4 所示，由于数据库收录论文存在时滞性（会议录和学位论文的收录时滞性更大），因而近几年发表的论文收录不全。从 2001～2012 年来看，海洋研究领域 EI 论文呈现快速增长趋势，2012 年论文数是 2001 年的 5 倍以上。

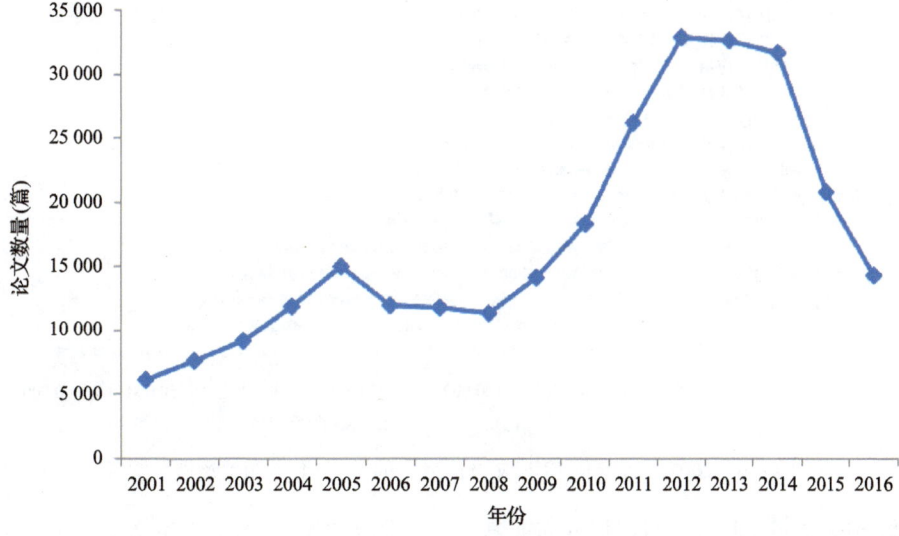

图 6-4　2001～2016 年海洋研究领域全球 EI 论文数量年度变化

图6-5统计了海洋学领域论文产出最多的15个机构的发文量。中国科学院EI论文产出数量位居全球首位，此外，哈尔滨工程大学、大连理工大学和中国海洋大学等3个机构也进入全球发文最多的前15个机构中。

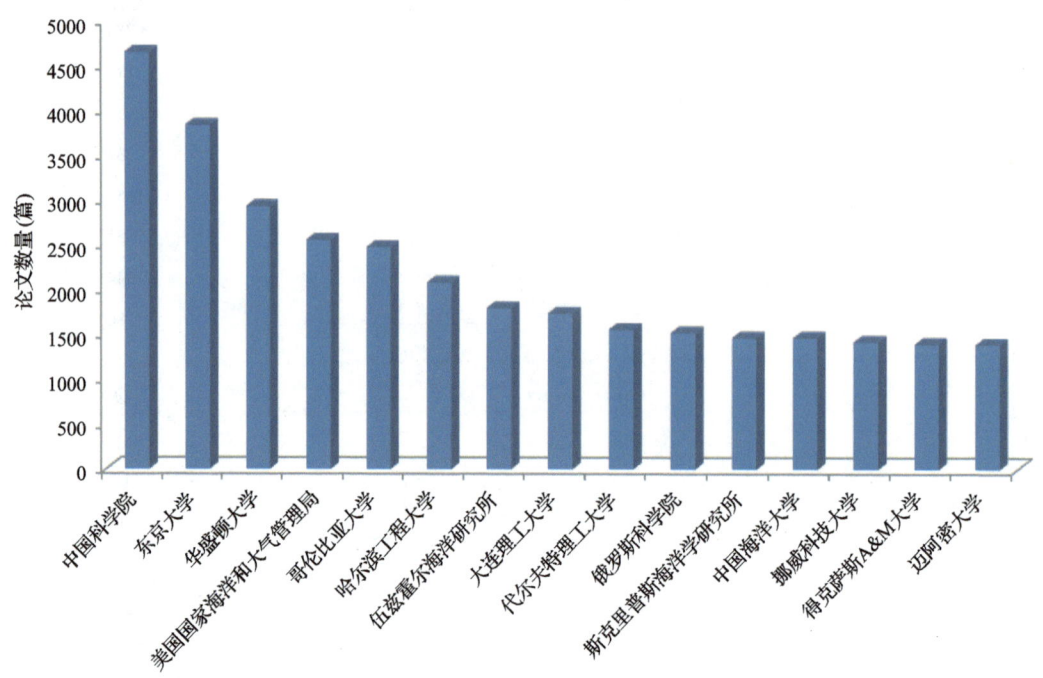

图6-5　2001～2016年全球EI论文产出最多的15个机构的发文量

图6-6统计了海洋相关主题分类领域中EI论文数最多的15个类目，主要分布在舰艇，海洋学总论，海水、潮汐和波浪，海洋科学与海洋学等领域。从学科领域分布来看，大量研究与数学、材料科学、力学、化学、生物工程与生物学、计算机应用等有关。

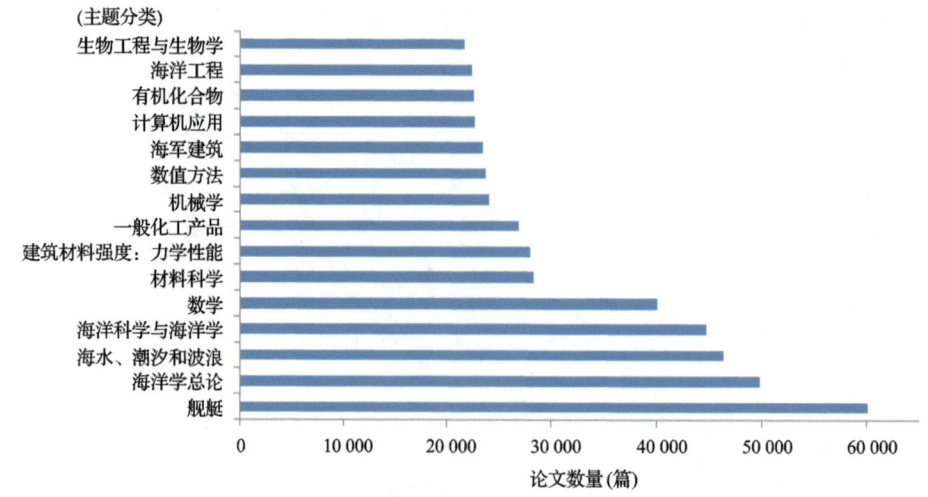

图6-6　2001～2016年EI论文产出最多的15个海洋相关领域发文量

发表海洋相关领域EI论文的期刊分布非常广泛。图6-7统计了发表论文最多的15个期刊，这15个期刊发表的论文数仅占海洋相关论文总数的9.94%。

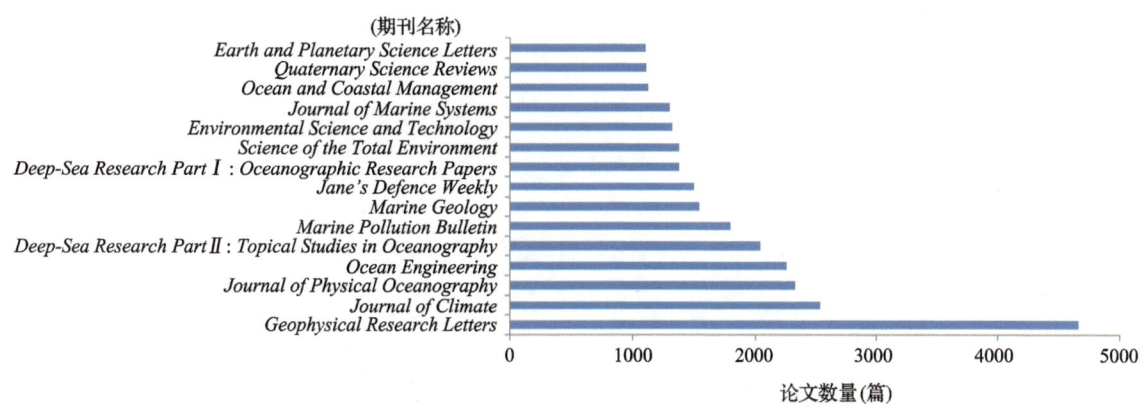

图 6-7　2001～2016 年海洋相关领域发表 EI 论文最多的期刊的发文量

会议和会议 EI 论文是了解领域国内外研究进展的重要渠道。图 6-8 统计了收录海洋相关论文最多的 15 个会议录的发文数。以海洋为主题的国际会议主要有：International Offshore and Polar Engineering Conference、International Conference on Ocean, Offshore and Arctic Engineering、International Conference on Port and Ocean Engineering under Arctic Conditions、the ISOPE Ocean Mining Symposium。此外，还有一些国家和地区会议，如 Annual Offshore Technology Conference、the Coastal Engineering Conference 等。

图 6-8　2001～2016 年发表 EI 论文最多的会议录的发文量

第二节　国家实力对比分析

一、基于 SCI 论文的分析

据 WOS 数据库 2001～2016 年论文统计总量，遴选全球海洋领域 SCI 发文量前 15 位的国家，如图 6-9 所示。美国占据绝对优势，其次为中国和英国，发文数量均在 9500 篇以上，前 15 位的其他国家分别为澳大利亚、法国、德国、加拿大、日本、西班牙、俄罗斯、意大利、挪威、荷兰、印度和韩国。

图 6-10 为 2001～2016 年海洋领域 SCI 发文量前 15 的国家年度发文情况，美国呈现稳定增长趋势，中国呈现明显增加趋势，尤其是最近 3 年（2014～2016 年），英国、澳大利亚、法国和德国的趋势相对稳定。

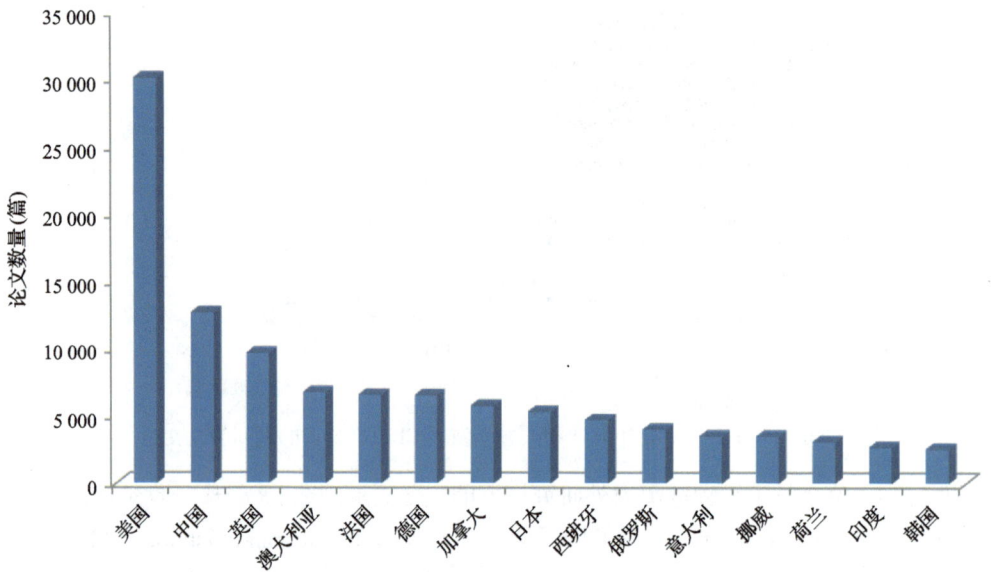

图 6-9　2001～2016 年全球海洋科技领域 SCI 论文前 15 的国家的发文量

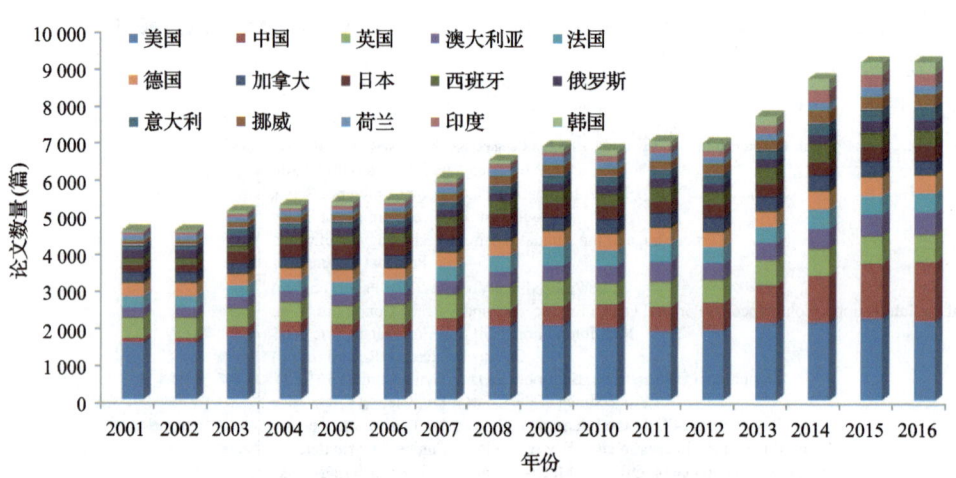

图 6-10　2001～2016 年海洋领域 SCI 论文前 15 的国家的年度发文量

以 WOS 数据库为数据源，统计全球海洋领域 SCI 发文数量前 15 国家的科研影响力及产出效率，包括论文总被引频次、篇均被引频次、未被引用论文数量及其占比及 H 指数、近 3 年发文量及其占比，如表 6-2 所示。

表 6-2　2001～2016 年全球海洋科技论文的影响力及产出效率指标统计

序号	国家	论文数量(篇)	篇均被引频次(次/篇)	总被引频次(次/篇)	近 3 年发文量(篇)	近 3 年发文占比(%)	未被引用论文数量(篇)	未被引用论文占比(%)	H 指数
1	美国	30 104	24.90	749 591	6 424	21	1 794	6	203
2	中国	12 672	9.00	114 098	4 282	34	2 132	17	190
3	英国	9 670	22.88	221 292	2 179	23	416	4	132
4	澳大利亚	6 769	21.67	146 653	1 758	26	264	4	116
5	法国	6 585	24.26	159 731	1 547	23	201	3	118
6	德国	6 555	24.45	160 271	1 533	23	243	4	129
7	加拿大	5 772	22.97	132 575	1 236	21	293	5	115

续表

序号	国家	论文数量(篇)	篇均被引频次(次/篇)	总被引频次(次/篇)	近3年发文量(篇)	近3年发文占比(%)	未被引用论文数量(篇)	未被引用论文占比(%)	H指数
8	日本	5 338	16.96	90 506	1 115	21	376	7	95
9	西班牙	4 735	20.30	96 109	1 254	26	195	4	96
10	俄罗斯	4 028	7.77	31 291	870	22	809	20	59
11	意大利	3 524	20.85	73 467	969	27	185	5	91
12	挪威	3 520	20.52	72 229	973	28	197	6	94
13	荷兰	3 118	25.96	80 945	710	23	111	4	103
14	印度	2 698	7.79	21 025	1 035	38	654	24	51
15	韩国	2 577	9.33	24 035	966	37	415	16	53

从主要国家科研影响力来看，论文总被引频次最高的为美国，其次为英国、德国、法国、澳大利亚、加拿大。我国尽管论文总量排名第二位，但是论文总被引频次却排名第七位。从主要国家海洋领域的SCI论文篇均被引频次来看，荷兰最高，约为26次/篇，美国、德国、法国均为24次/篇以上，中国仅为9次/篇。被引频次偏低一般归因于论文影响力不足或者论文多为近几年发文造成影响周期滞后等。例如，中国近3年SCI发文数量占全球比例为34%，这可能是篇均被引频次较低的一个重要原因。从未被引用论文数量来看，中国最多；从未被引用论文占比来看，法国最少，约为3%，其次为英国、澳大利亚、德国、西班牙和荷兰，中国为17%。

H指数可用于评估一个国家的科研论文影响力，因为H指数同时关注论文被引数量和被引频次指标，H指数与总被引频次、论文被引数量具有较强的正相关关系。国家H指数，主要指在一个国家发表的Np篇论文中，如果有H篇论文的被引频次都大于等于H，而其他$(Np-H)$篇论文被引频次都小于H，那么此国家的科研成就的指数值为H。在前15的国家中，美国、中国、英国和德国的H指数较高，表明这些国家在海洋领域中的科研成就较为突出。

从主要国家的科技产出效率指标来看，最近3年发表论文较多的国家为美国、中国、英国和澳大利亚。从近3年发文数量占据所有统计年份数量的比例来看，中国、印度和韩国近3年均超过了30%，表明这些国家海洋创新正在崛起。

二、基于EI论文的分析

2001～2016年全球海洋领域发表EI论文最多的10个国家如图6-11所示。美国最多，中国紧随其后，两国论文数量占全球论文数量的40%左右，是海洋学EI论文产出最主要的2个国家。

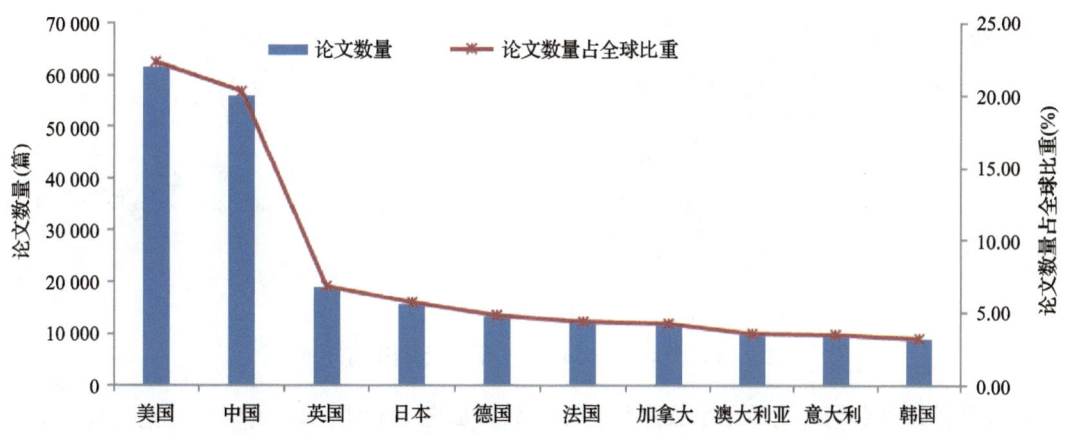

图6-11　2001～2016年全球海洋领域发表EI论文最多的10个国家的论文数量及其占全球的比重

图 6-12 统计了海洋领域 EI 论文数量前 10 位国家的年度变化，中国和美国近年来的年度增长幅度远高于其他国家。2011 年以来，中国论文数量增长尤其迅速，产出量已超过美国，位居全球首位。

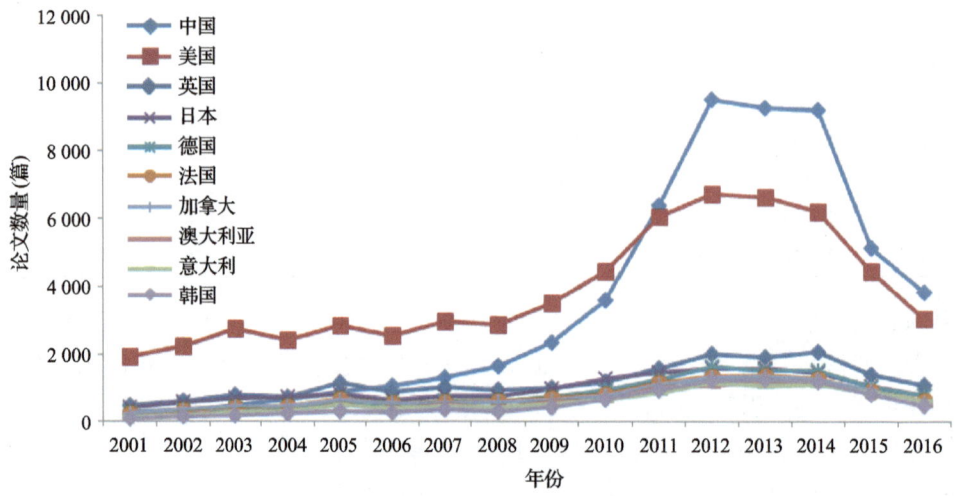

图 6-12　2001～2016 年海洋领域发表 EI 论文最多的 10 个国家及其论文数量年度变化趋势

第三节　海洋领域专利技术成果分析

一、国际总体研发格局

以 DII（Derwent Innovations Index）数据库检索 2001～2016 年海洋专利申请数据，如图 6-13 所示，中国专利申请数量位居第一，遥遥领先于其他国家、地区与专利组织，与韩国、日本、美国和世界知识产权组织国际局的合计数量相当。

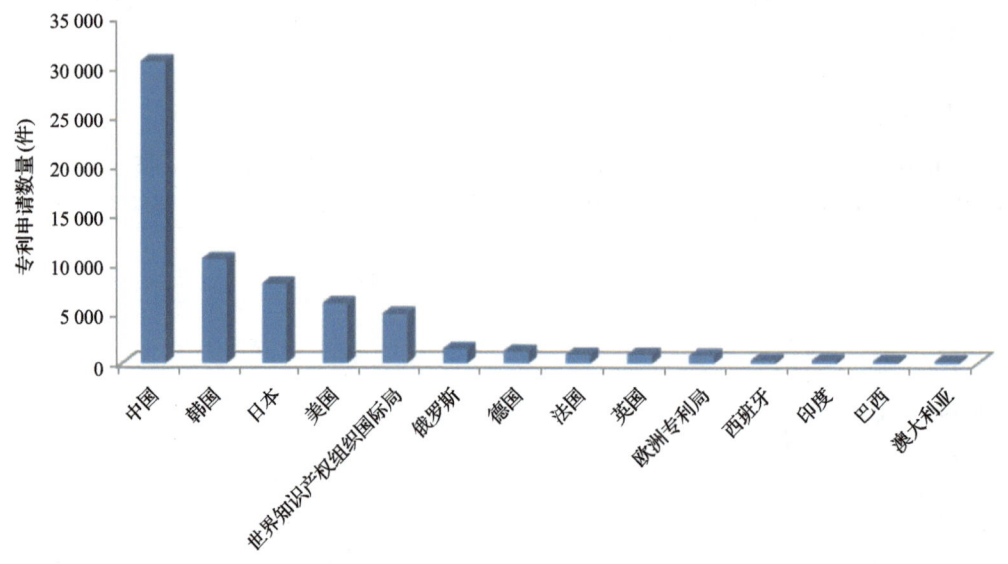

图 6-13　2001～2016 年全球海洋专利申请分布

中国与世界海洋专利申请数量增长态势基本一致，如图 6-14 所示，占世界的比例稳步提高，从 2001 年的 5.4% 提高至 2016 年 84.4%（包含多国合作专利申请）。

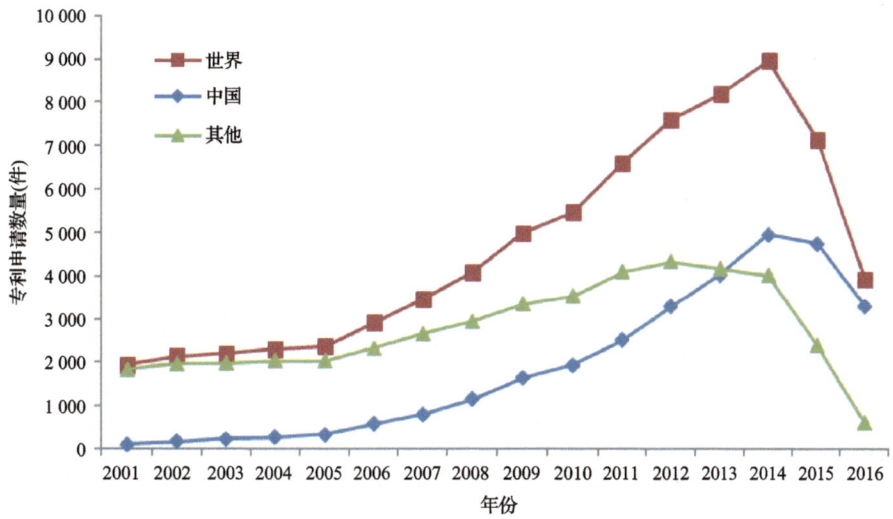

图 6-14　2001~2016 年世界海洋专利申请数量年度变化

世界海洋专利申请人逐年增多,如图 6-15 所示,由 2001 年的不足 1000 人增加到 2015 年的 5757 人,海洋产业参与人数增量接近 5 倍,表明海洋产业逐渐壮大。

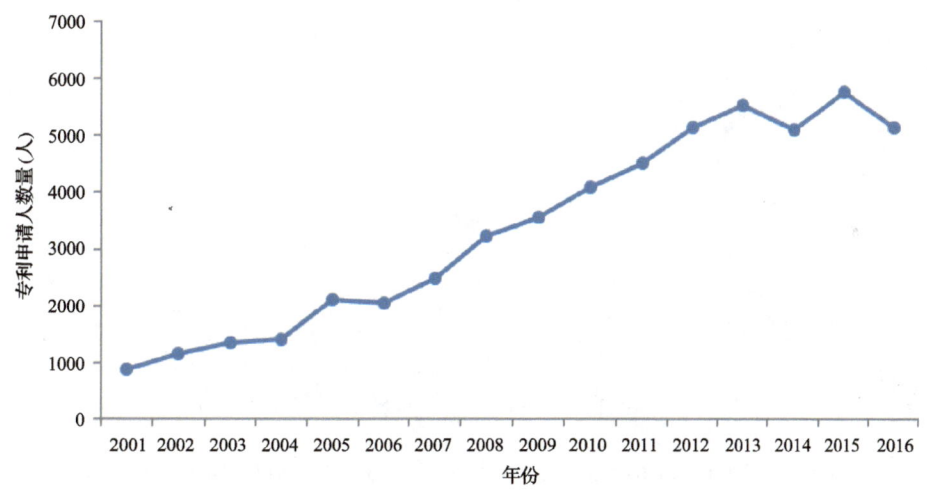

图 6-15　2001~2016 年世界海洋专利申请人数量年度变化

世界海洋专利申请机构也逐年增多,如图 6-16 所示,由 2001 年的 949 家增长至 2016 年的 3178 家。
2001~2016 年世界海洋专利申请前 15 位的机构如图 6-17 所示,中国有 5 家机构位列前 15 位中,分别是中国海洋石油集团有限公司、浙江海洋大学、中国海洋大学、浙江大学和大连海洋大学。

如图 6-18 所示,世界海洋专利按照 IPC 分类前 15 位的专利分别是:B63B(船舶或其他水上船只;船用设备)、C02F(污水、污泥污染处理)、A01K(畜牧业;禽类、鱼类、昆虫的管理;捕鱼;饲养或养殖其他类不包含的动物;动物的新品种)、A23L(不包含在 A21D 或 A23B~A23J 小类中的食品、食料或非酒精饮料)、A61K(医学用配置品)、E02B(水利工程)、F03B(液力机械或液力发动机)、B01D(分离)、A61P(化合物或药物制剂的治疗活性)、B63H(船舶的推进装置或操舵装置)、E21B(土层或岩石的钻进)、G01N(借助测定材料的化学或者物理性质来测试或分析材料)、G01V(地球物理;重力测量;物质或物体的探测;示踪物)、E02D(基础;挖方;填方;地下或水下结构物)、G01S(无线电定向;无线电导航;采用无线电波测距或测速;采用无线电波的反射或再辐射的定位或存在检测;采用其他波的类似装置)。

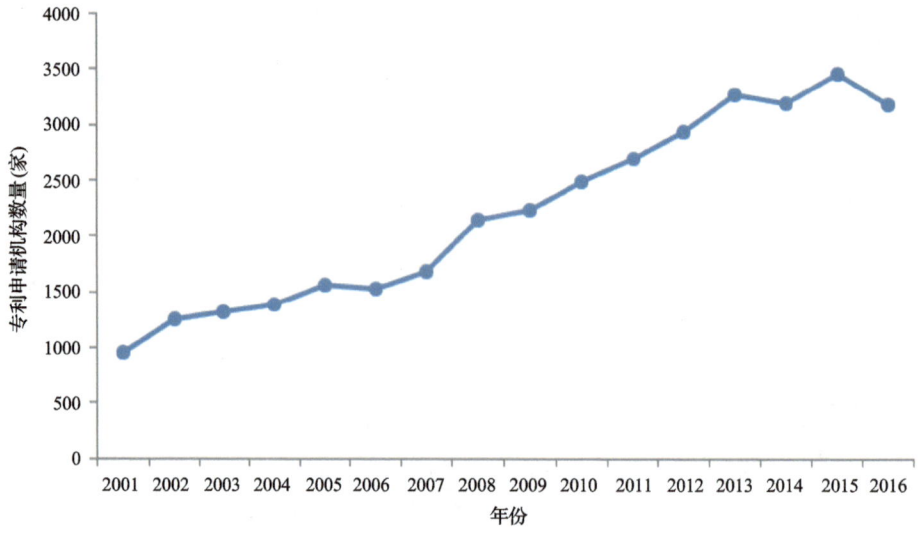

图 6-16 2001~2016 年世界海洋专利申请机构数量年度变化

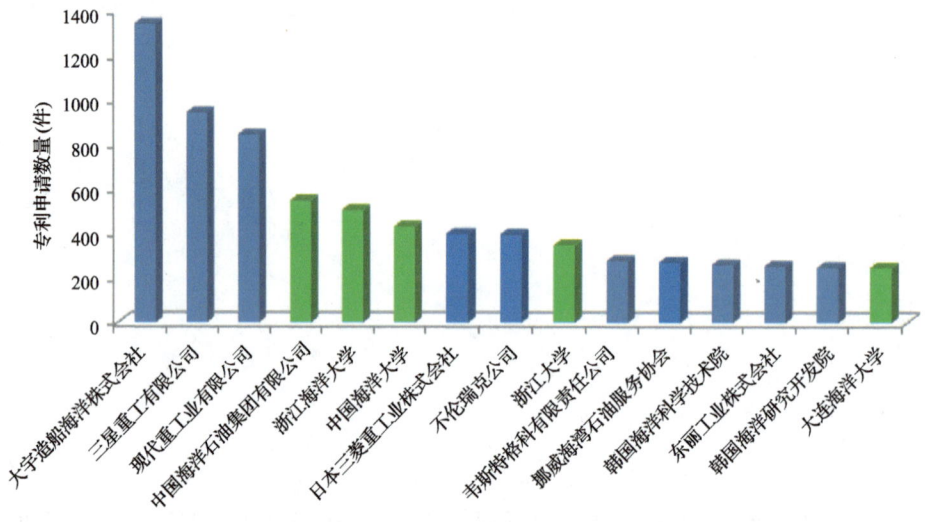

图 6-17 2001~2016 年世界海洋专利申请前 15 位的机构专利申请数量比较

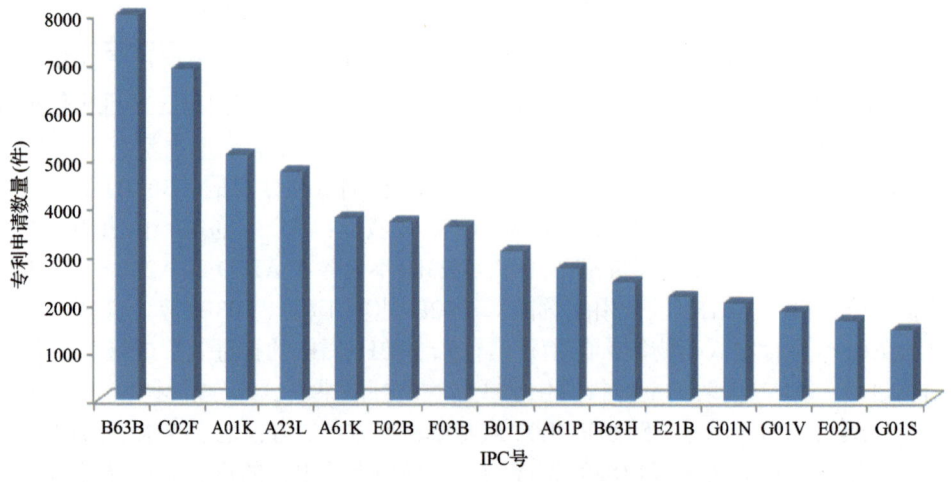

图 6-18 2001~2016 年世界海洋主要国际专利分类号(IPC 号)的专利申请数量比较

二、国家技术研发实力比较

对专利申请量排名前 14 位的国家和组织进行比较分析,发现中国专利申请数量增长优势明显,如图 6-19 所示。中国自 2006 年专利申请数量飞速上升以来,一直位于世界前列,并且超出其他国家的数量越来越多。这表明中国近年来海洋产业逐渐壮大,其可能与国家专利扶持政策相关。

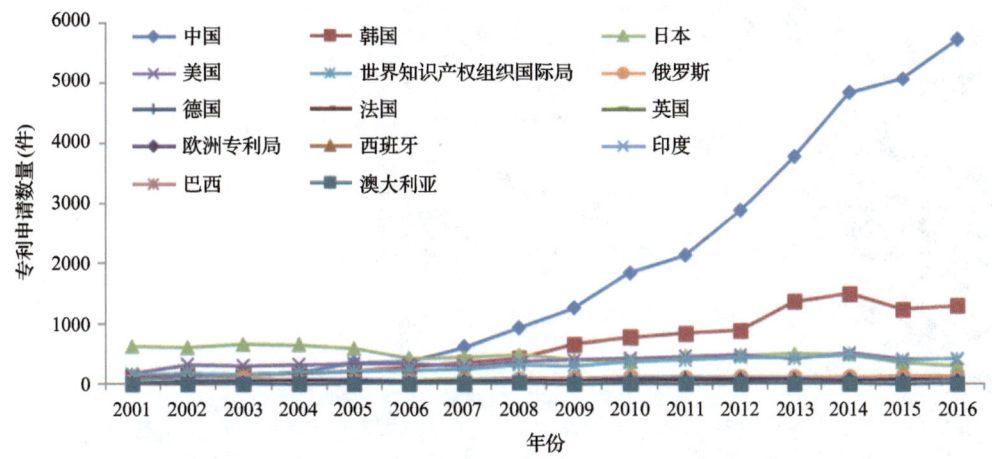

图 6-19　2001～2016 年专利申请量排名前 14 的国家和组织年度专利申请数量变化

在近 3 年专利申请数量占比中,中国占比超过一半,如图 6-20 所示,随后是韩国和印度,占比在 38%左右。

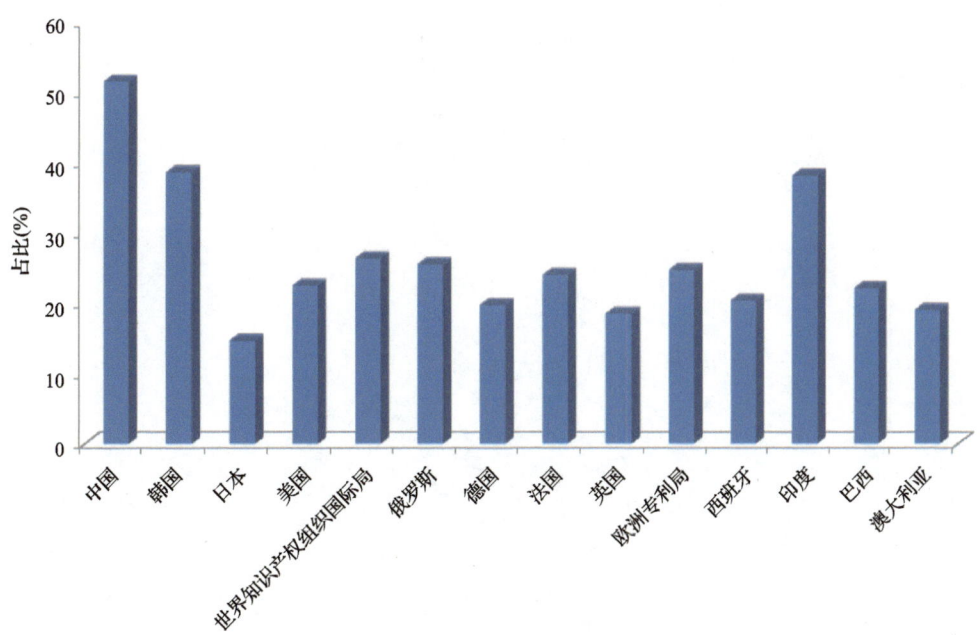

图 6-20　2001～2016 年专利申请量排名前 14 的国家和组织近 3 年专利申请数量占比

附 录

附录一 国家海洋创新指数指标体系

一、国家海洋创新指数内涵

国家海洋创新指数是指衡量一国海洋创新能力，切实反映一国海洋创新质量和效率的综合性指数。

国家海洋创新指数评价工作借鉴了国内外关于国家竞争力和创新评价等理论与方法，基于创新型海洋强国的内涵分析，确定指标选择原则，从海洋创新资源、海洋知识创造、海洋创新绩效和海洋创新环境 4 个方面构建了国家海洋创新指数的指标体系，力求全面、客观、准确地反映我国海洋创新能力在创新链不同层面的特点，形成一套比较完整的指标体系和评价方法。通过指数测度，为综合评价创新型海洋强国建设进程、完善海洋创新政策提供技术支撑和咨询服务。

二、创新型海洋强国内涵

建设海洋强国，亟需推动海洋科技向创新引领型转变。国际历史经验表明，海洋科技发展是实现海洋强国的根本保障，应建立国家海洋创新评价指标体系，从战略高度审视我国海洋发展动态，强化海洋基础研究和人才团队建设，大力发展海洋科学技术，为经济社会各方面提供决策支持。

国家海洋创新指数评价将有利于国家和地方政府及时掌握海洋科技发展战略实施进展及可能出现的问题，为进一步采取对策提供基本信息；有利于国外、国内公众了解我国海洋事业取得的进展、成就、趋势及存在的问题；有利于企业和投资者研判我国海洋领域的机遇与风险；有利于为从事海洋领域研究的学者和机构提供有关信息。

纵观我国海洋经济的发展历程，大体经历了三个阶段：资源依赖阶段、产业规模粗放扩张阶段和由量向质转变阶段。海洋科技的飞速发展，推动了新型海洋产业规模不断发展扩大，成为海洋经济新的增长点。我国海域辽阔、海洋资源丰富，但是多年的粗放式发展使得资源环境问题日益突出，制约了海洋经济的进一步发展。因此，只有不断地进行海洋创新，才能促进海洋经济的健康发展，使我国步入创新型海洋强国行列。

创新型海洋强国最主要的特征是国家海洋经济社会发展方式与传统的发展模式相比发生了根本的变化。创新型海洋强国的判别应主要依据海洋经济增长是主要依靠要素(传统的海洋资源消耗和资本)投入来驱动，还是主要依靠以知识创造、传播和应用为标志的创新活动来驱动。

创新型海洋强国应具备 4 个方面的能力：①较高的海洋创新资源综合投入能力；②较高的海洋知识创造与扩散应用能力；③较高的海洋创新绩效影响表现能力；④良好的海洋创新环境。

三、指标选择原则

(1)评价思路体现海洋可持续发展思想。不仅要考虑海洋创新整体发展环境，还要考虑经济发展、知识成果的可持续性指标，兼顾指数的时间趋势。

(2)数据来源具有权威性。基本数据必须来源于公认的国家官方统计和调查。通过正规渠道定期搜集，确保基本数据的准确性、权威性、持续性和及时性。

(3)指标具有科学性、现实性和可扩展性。海洋创新指数与各项分指数之间逻辑关系严密，分指数的每一个指标都能体现科学性和客观性思想,尽可能减少人为合成指标,各指标均有独特的宏观表征意义,定义相对宽泛,并非对应唯一狭义数据,便于指标体系的扩展和调整。

(4) 评价体系兼顾我国海洋区域特点。选取指标以相对指标为主,兼顾不同区域在海洋创新资源产出效率、创新活动规模和创新领域广度上的不同特点。

(5) 纵向分析与横向比较相结合。既有纵向的历史发展轨迹回顾分析,也有横向的各沿海区域、各经济区、各经济圈比较和国际比较。

四、指标体系构建

创新是从创新概念提出到研发、知识产出再到商业化应用转化为经济效益的完整过程。海洋创新能力体现在海洋科技知识的产生、流动和转化为经济效益的整个过程中。应该从海洋创新环境、创新资源的投入、知识创造与应用、绩效影响等整个创新链的主要环节来构建指标,评价国家海洋创新能力。

本报告采用综合指数评价方法,从创新过程选择分指数,确定了海洋创新资源、海洋知识创造、海洋创新绩效和海洋创新环境 4 个分指数;遵循指标的选取原则,选择 20 个指标(附表 1-1)形成国家海洋创新指数评价指标体系,指标均为正向指标;再利用国家海洋创新综合指数及其指标体系对我国海洋创新能力进行综合分析、比较与判断。

附表 1-1 国家海洋创新指数指标体系

综合指数	分指数	指标
国家海洋创新指数 A	海洋创新资源 B_1	1.研究与发展经费投入强度 C_1
		2.研究与发展人力投入强度 C_2
		3.R&D 人员中博士占比 C_3
		4.科技活动人员占海洋科研机构从业人员的比重 C_4
		5.万名科研人员承担的课题数 C_5
	海洋知识创造 B_2	6.亿美元经济产出的发明专利申请数 C_6
		7.万名 R&D 人员的发明专利授权数 C_7
		8.本年出版科技著作 C_8
		9.万名科研人员发表的科技论文数 C_9
		10.国外发表的论文数占总论文数的比重 C_{10}
	海洋创新绩效 B_3	11.海洋科技成果转化率 C_{11}
		12.海洋科技进步贡献率 C_{12}
		13.海洋劳动生产率 C_{13}
		14.科研教育管理服务业占海洋生产总值的比重 C_{14}
		15.单位能耗的海洋经济产出 C_{15}
		16.海洋生产总值占国内生产总值的比重 C_{16}
	海洋创新环境 B_4	17.沿海地区人均海洋生产总值 C_{17}
		18.R&D 经费中设备购置费所占比重 C_{18}
		19.海洋科研机构科技经费筹集额中政府资金所占比重 C_{19}
		20.R&D 人员人均折合全时工作量 C_{20}

海洋创新资源: 反映一个国家海洋创新活动的投入力度、创新型人才资源供给能力及创新所依

赖的基础设施投入水平。创新投入是国家海洋创新活动的必要条件,包括科技资金投入和人才资源投入等。

海洋知识创造:反映一个国家的海洋科研产出能力和知识传播能力。海洋知识创造的形式多种多样,产生的效益也是多方面的,本报告主要从海洋发明专利和科技论文等角度考虑海洋创新的知识积累效益。

海洋创新绩效:反映一个国家开展海洋创新活动所产生的效果和影响。海洋创新绩效分指数从国家海洋创新的效率和效果两个方面选取指标。

海洋创新环境:反映一个国家海洋创新活动所依赖的外部环境,主要包括相关海洋制度创新和环境创新。其中,制度创新的主体是政府等相关部门,主要体现在政府对创新的政策支持、对创新的资金支持和知识产权管理等方面;环境创新主要指创新的配置能力、创新基础设施、创新基础经济水平、创新金融及文化环境等。

附录二 国家海洋创新指数指标解释

C_1. 研究与发展经费投入强度

海洋科研机构的 R&D 经费占国内海洋生产总值比重,也就是国家海洋研发经费投入强度指标,反映国家海洋创新资金投入强度。

C_2. 研究与发展人力投入强度

每万名涉海就业人员中 R&D 人员数,反映一个国家创新人力资源投入强度。

C_3. R&D 人员中博士占比

海洋科研机构内 R&D 人员中博士毕业人员所占比重,反映一个国家海洋科技活动的顶尖人才力量。

C_4. 科技活动人员占海洋科研机构从业人员的比重

海洋科研机构内从业人员中科技活动人员所占比重,反映一个国家海洋创新活动科研力量的强度。

C_5. 万名科研人员承担的课题数

平均每万名科研人员承担的国内课题数,反映海洋科研人员从事创新活动的强度。

C_6. 亿美元经济产出的发明专利申请数

一国海洋发明专利申请数量除以海洋生产总值(以汇率折算的亿美元为单位)。该指标反映了相对于经济产出的技术产出量和一个国家的海洋创新活动的活跃程度。3 种专利(发明专利、实用新型专利和外观设计专利)中发明专利技术含量和价值最高,发明专利申请数可以反映一个国家的海洋创新活动的活跃程度和自主创新能力。

C_7. 万名 R&D 人员的发明专利授权数

平均每万名 R&D 人员的国内发明专利授权量,反映一个国家的自主创新能力和技术创新能力。

C_8. 本年出版科技著作

指经过正式出版部门编印出版的科技专著、大专院校教科书、科普著作。只统计本单位科技人员为第一作者的著作，同一书名计为一种著作，与书的发行量无关，反映一个国家海洋科学研究的产出能力。

C_9. 万名科研人员发表的科技论文数

平均每万名科研人员发表的科技论文数，反映科学研究的产出效率。

C_{10}. 国外发表的论文数占总论文数的比重

一国发表的科技论文中，在国外发表的论文所占比重，可反映科技论文相关研究的国际化水平。

C_{11}. 海洋科技成果转化率

衡量海洋科技创新成果转化为商业开发产品的指数，是指为提高生产力水平而对科学研究与技术开发所产生的具有实用价值的海洋科技成果所进行的后续试验、开发、应用、推广直至形成新产品、新工艺、新材料，发展新产业等活动占海洋科技成果总量的比值。

C_{12}. 海洋科技进步贡献率

海洋科技进步贡献率的定义应以海洋科技进步增长率的定义为基础，是指在海洋经济各行业中，海洋科技进步增长率在整个海洋经济增长率中所占的比例。而海洋科技进步增长率则是指人类利用海洋资源和海洋空间进行各类社会生产、交换、分配和消费等活动时，剔除资金和劳动等生产要素以外其他要素的增长，具体是指由技术创新、技术扩散、技术转移与引进引起的装备技术水平的提高、技术工艺的改良、劳动者素质的提升及管理决策能力的增强等。

C_{13}. 海洋劳动生产率

采用涉海就业人员的人均海洋生产总值，反映海洋创新活动对海洋经济产出的作用。

C_{14}. 科研教育管理服务业占海洋生产总值的比重

反映海洋科研、教育、管理及服务等活动对海洋经济的贡献程度。

C_{15}. 单位能耗的海洋经济产出

采用万吨标准煤能源消耗的海洋生产总值，用来测度海洋创新带来的减少资源消耗的效果，也反映一个国家海洋经济增长的集约化水平。

C_{16}. 海洋生产总值占国内生产总值的比重

反映海洋经济对国民经济的贡献，用来测度海洋创新对海洋经济的推动作用。

C_{17}. 沿海地区人均海洋生产总值

按沿海地区人口平均的海洋生产总值，它在一定程度上反映了沿海地区人民的生活水平，可以衡量海洋生产力的增长情况和海洋创新活动所处的外部环境。

C_{18}. R&D 经费中设备购置费所占比重

海洋科研机构的 R&D 经费中设备购置费所占比重，反映海洋创新所需的硬件设备条件，在一定程度上反映了海洋创新的硬环境。

C_{19}. 海洋科研机构科技经费筹集额中政府资金所占比重

反映政府投资对海洋创新的促进作用及海洋创新所处的制度环境。

C_{20}. R&D 人员人均折合全时工作量

反映一个国家海洋科技人力资源投入工作量与全时工作能力。

附录三 国家海洋创新指数评价方法

国家海洋创新指数的计算方法采用国际上流行的标杆分析法，即洛桑国际竞争力评价采用的方法。标杆分析法是目前国际上广泛采用的一种评价方法，其原理是：对被评价的对象给出一个基准值，并以此标准去衡量所有被评价的对象，从而发现彼此之间的差距，给出排序结果。

采用海洋创新评价指标体系中的指标，利用 2002~2016 年指标数据，分别计算以后各年的海洋创新指数与分指数得分，与基年比较即可看出国家海洋创新指数增长情况。

一、原始数据标准化处理

设定 2004 年为基准年，基准值为 100。对国家海洋创新指数指标体系中 20 个指标的原始值进行标准化处理。具体操作为

$$C_j^t = \frac{100 x_j^t}{x_j^1}$$

式中，$j=1\sim20$，为指标序列编号；$t=1\sim13$，为 2004~2016 年编号；x_j^t 表示各年各项指标的原始数据值（x_j^1 表示 2004 年各项指标的原始数据值）；C_j^t 表示各年各项指标标准化处理后的值。

二、国家海洋创新分指数测算

采用等权重[①]（下同）测算各年国家海洋创新指数分指数得分。

当 $i=1$ 时，$B_1^t = \sum_{j=1}^{5} \beta_1 C_j^t$，其中 $\beta_1 = \frac{1}{5}$；

当 $i=2$ 时，$B_2^t = \sum_{j=6}^{10} \beta_2 C_j^t$，其中 $\beta_2 = \frac{1}{5}$；

当 $i=3$ 时，$B_3^t = \sum_{j=11}^{16} \beta_3 C_j^t$，其中 $\beta_3 = \frac{1}{6}$；

当 $i=4$ 时，$B_4^t = \sum_{j=17}^{20} \beta_4 C_j^t$，其中 $\beta_4 = \frac{1}{4}$。

式中，β 为权重，$t=1\sim13$，i 代表分指数，B_1^t、B_2^t、B_3^t、B_4^t 依次代表各年海洋创新资源分指数、海洋知识创造分指数、海洋创新绩效分指数和海洋创新环境分指数的得分。

---

① 采用《国家海洋创新指数报告 2016》的权重选取方法，取等权重

三、国家海洋创新指数测算

采用等权重(同上)测算国家海洋创新指数得分,即

$$A^t = \sum_{i=1}^{4} \varpi B_i^t$$

式中,$i=1\sim4$;$t=1\sim13$,为2004~2016年编号;ϖ为权重(等权重为$\frac{1}{4}$);A^t为各年的国家海洋创新指数得分。

附录四 区域海洋创新指数评价方法

一、区域海洋创新指数指标体系说明

区域海洋创新指数由海洋创新资源、海洋知识创造、海洋创新绩效和海洋创新环境4个分指数构成。与国家海洋创新指数指标体系相比,区域海洋创新绩效分指数相比于国家海洋创新绩效分指数缺少"海洋科技进步贡献率"和"海洋科技成果转化率"2个指标。

二、原始数据归一化处理

对2016年18个指标的原始值分别进行归一化处理。归一化处理是为了消除多指标综合评价中计量单位的差异和指标数值的数量级、相对数形式的差别,解决数据指标的可比性问题,使各指标处于同一数量级,便于进行综合对比分析。

指标数据处理采用直线型归一化方法,即

$$c_{ij} = \frac{y_{ij} - \min y_{ij}}{\max y_{ij} - \min y_{ij}}$$

式中,$i=1\sim11$,为我国大陆11个沿海省(自治区、直辖市)序列号;$j=1\sim18$,为指标序列号;y_{ij}表示各项指标的原始数据值;c_{ij}表示各项指标归一化处理后的值。

三、区域海洋创新分指数计算

区域海洋创新资源分指数得分 $b_1 = 100 \times \sum_{j=1}^{5} \varphi_1 c_j$,其中 $\varphi_1 = \frac{1}{5}$;

区域海洋知识创造分指数得分 $b_2 = 100 \times \sum_{j=6}^{10} \varphi_2 c_j$,其中 $\varphi_2 = \frac{1}{5}$;

区域海洋创新绩效分指数得分 $b_3 = 100 \times \sum_{j=11}^{14} \varphi_3 c_j$,其中 $\varphi_3 = \frac{1}{4}$;

区域海洋创新环境分指数得分 $b_4 = 100 \times \sum_{j=15}^{18} \varphi_4 c_j$,其中 $\varphi_4 = \frac{1}{4}$。

式中,$j=1\sim18$,b_1、b_2、b_3、b_4依次代表区域海洋创新资源分指数、区域海洋知识创造分指数、区域海洋创新绩效分指数和区域海洋创新环境分指数的得分。

四、区域海洋创新指数计算

采用等权重（同国家海洋创新指数）测算区域海洋创新指数得分。

$$a = \frac{1}{4}(b_1 + b_2 + b_3 + b_4)$$

式中，a 为区域海洋创新指数得分。

附录五 海洋科技进步贡献率测算方法

目前，进行科技进步贡献率测算广泛而常用的方法是索洛余值法，这也是国家发展和改革委员会（原国家计划委员会）、国家统计局及科学技术部等系统普遍使用的方法。

索洛余值法以科布-道格拉斯生产函数作为基础模型，该方法表明了经济增长除取决于资本增长率、劳动增长率及资本和劳动对收入增长的相对作用的权数以外，还取决于技术进步，区分了由要素数量增加而产生的"增长效应"和因要素技术水平提高而带来经济增长的"水平效应"，系统地解释了经济增长的原因。

海洋经济涉及多个行业和部门，为了综合反映海洋类各行业的科技进步对海洋经济整体增长的贡献，需要对海洋类各行业进行全面测算，再按照各行业经济总产值在海洋经济整体中所占的比重，将各行业的科技进步在增长速度测算阶段进行汇总加权，得出海洋科技进步增长率，并进一步测算得出海洋科技进步贡献率。

根据海洋科技进步贡献率的理论内涵和特点，海洋科技进步贡献率可涉及的海洋产业范围有：直接从海洋中获取产品的生产和服务；直接对从海洋中获取的产品所进行的一次性加工生产和服务；直接应用于海洋的产品生产和服务；利用海水或海洋空间作为生产过程的基本要素所进行的生产和服务。其中，海洋科学研究、教育、技术等其他服务和管理范畴不适宜纳入海洋科技进步贡献率测算范围。

结合我国海洋科技的特点，通过对 8 个海洋产业的产出增长率、资本增长率和劳动增长率进行行业加权，构建海洋科技进步贡献率测算的基本公式，公式推导过程如下。

令第 i 个产业（$i=1,2,3,\cdots,8$）分别代表海洋养殖业、海洋捕捞业、海洋盐业、海洋船舶工业、海洋石油业、海洋天然气产业、海洋交通运输业、滨海旅游业；$y_i(t)$ 表示第 i 个产业 t 期的产出增长率，其中 $t \in [t_1, t_2]$；$k_i(t)$ 与 $l_i(t)$ 分别表示 t 期的资本与劳动投入增长率，其中 $t \in [t_1, t_2]$；γ_i 代表第 i 个产业在总海洋产业中的权重。k_i、l_i、y_i 分别表示 $k_i(t)$、$l_i(t)$、$y_i(t)$ 研究区间 t_1 至 t_2 内的平均值，即

$$k_i = \frac{\sum_{t=t_1}^{t_2} k_i(t)}{n}, \quad l_i = \frac{\sum_{t=t_1}^{t_2} l_i(t)}{n}, \quad y_i = \frac{\sum_{t=t_1}^{t_2} y_i(t)}{n} \quad (n = t_2 - t_1)$$

$$k = \sum_{i=1}^{8} k_i \gamma_i, \quad l = \sum_{i=1}^{8} l_i \gamma_i, \quad y = \sum_{i=1}^{8} y_i \gamma_i$$

式中，k、l、y 分别表示 k_i、l_i、y_i 的加权平均值，由此可得

$$A = 1 - \frac{\alpha k}{y} - \frac{\beta l}{y} = 1 - \frac{\alpha \sum_{i=1}^{8} k_i \gamma_i}{\sum_{i=1}^{8} y_i \gamma_i} - \frac{\beta \sum_{i=1}^{8} l_i \gamma_i}{\sum_{i=1}^{8} y_i \gamma_i}$$

$$= 1 - \frac{\alpha \sum_{i=1}^{8} \frac{\sum_{t=t_1}^{t_2} k_i(t)}{n} \gamma_i}{\sum_{i=1}^{8} \frac{\sum_{t=t_1}^{t_2} y_i(t)}{n} \gamma_i} - \frac{\beta \sum_{i=1}^{8} \frac{\sum_{t=t_1}^{t_2} l_i(t)}{n} \gamma_i}{\sum_{i=1}^{8} \frac{\sum_{t=t_1}^{t_2} y_i(t)}{n} \gamma_i}$$

式中，A 表示研究期内的海洋科技进步贡献率；α 与 β 分别表示海洋产业资本和劳动的弹性系数。

在指标时长的选取方面，由于海洋科技对海洋经济的影响是长期的，海洋科技进步贡献率测算时间在 10 年以上为妥，最少 5 年。综合考虑海洋管理实际需要和海洋数据年限限制，本研究在"十一五"期间指标测算和"十二五"期间指标短期预测时使用 5 年数据平均值，其他测算和长期预测时使用 10 年数据平均值(根据 2006~2016 年时长而定)。

在海洋产业的选取上，根据《中国海洋统计年鉴 2017》，2016 年我国主要海洋产业包括海洋渔业(16.26%)、海洋油气业(3.06%)、海洋矿业(0.24%)、海洋盐业(0.14%)、海洋船舶工业(5.26%)、海洋化工业(3.39%)、海洋生物医药业(1.20%)、海洋工程建筑业(6.10%)、海洋电力业(0.45%)、海水利用业(0.05%)、海洋交通运输业(20.08%)和滨海旅游业(43.79%)12 大产业(附表 5-1)。经初步筛选和可行性分析，确定数据可支持的 8 个可测算行业包括：海水养殖业、海洋捕捞业、海洋盐业、海洋船舶工业、海洋石油业、海洋天然气产业、海洋交通运输业、滨海旅游业。以上 8 个海洋行业的产值总和约占主要海洋产业总值的 86.96%，基本能够有效地反映我国海洋经济发展情况。

附表 5-1　2016 年我国主要海洋产业增加值

主要海洋产业	增加值(亿元)	占比(%)
海洋渔业	4 615.4	16.26
海洋油气业	868.8	3.06
海洋矿业	67.3	0.24
海洋盐业	38.9	0.14
海洋船舶工业	1 492.4	5.26
海洋化工业	961.8	3.39
海洋生物医药业	341.3	1.20
海洋工程建筑业	1 731.3	6.10
海洋电力业	128.5	0.45
海水利用业	13.7	0.05
海洋交通运输业	5 699.8	20.08
滨海旅游业	12 432.8	43.79
合计	28 392	—

在弹性系数的确定方面，计算海洋科技进步贡献率时，可采用经验估计法、比值法和回归法确定资本和劳动产出弹性系数。经验估计法是指借鉴其他权威专家所测算出的系数；比值法的原理是利用与资本投入量和劳动投入量有关的数据计算两者的比值；回归法是指采用有约束（即 $\alpha+\beta=1$）或无约束的生产函数模型，代入相应数值后，根据计量方法（即利用最小二乘法进行回归）估算出两个弹性系数。本次测算采用的是 $\alpha=0.3$，$\beta=0.7$。

在权重的确定方面，根据《中国海洋统计年鉴》中我国"十二五"期间 8 个海洋行业的产值情况，确定各行业权重值（附表 5-2）。

附表 5-2　各产业权重值

产业	权重	产业	权重
海水养殖业	0.1096	海洋石油业	0.0709
海洋捕捞业	0.0810	海洋天然气产业	0.0045
海洋盐业	0.0003	海洋交通运输业	0.2489
海洋船舶工业	0.0664	滨海旅游业	0.4154

在数据来源方面，本研究使用的代表海洋产业产值、资本和劳动的指标数据均来源于相应年份的《中国海洋统计年鉴》（附表 5-3）。从数据基础来看，目前可用于测算的连续数据为 1996~2016 年海洋产业产值、资本和劳动数据（对个别缺失数据进行趋势拟合插值）。

附表 5-3　八大产业的产出、资本和劳动指标

产业	产出指标	资本指标	劳动指标
海水养殖业	海水养殖产量	海水养殖面积	海洋渔业及相关产业就业人员数
海洋捕捞业	海洋捕捞产量	主要海上活动船舶总吨	海洋渔业及相关产业就业人员数
海洋盐业	沿海地区海盐产量	盐业生产面积	海洋盐业就业人员数
海洋船舶工业	海洋船舶工业增加值	沿海地区造船完工量	海洋船舶工业就业人员数
海洋石油业	沿海地区海洋原油产量	海洋采油井	海洋石油和天然气业就业人员数
海洋天然气产业	沿海地区海洋天然气产量	海洋采气井	海洋石油和天然气业就业人员数
海洋交通运输业	海洋交通运输业增加值	沿海规模以上港口生产用码头泊位个数	海洋交通运输业就业人员数
滨海旅游业	滨海旅游业增加值	沿海地区旅行社总数	滨海旅游业就业人员数

将各行业的基准数据代入海洋科技进步贡献率公式，经调整和验证，得出我国"十一五"期间海洋科技进步贡献率的平均值为 54.4%，2006~2016 年海洋科技进步贡献率的平均值为 65.9%。

附录六　海洋科技成果转化率测算方法

海洋科技成果转化率的定义源于科技成果转化率。在科技成果转化率的研究方面，国外学者很少直接使用"科技成果转化"，而是用"科技经济一体化""技术创新""技术转化""技术推广""技术扩散"或"技术转移"来代替，且国外并没有针对全社会领域进行科技成果转化情况的统计或评价。

从国内来看，各领域学者对于科技成果转化率的定义不尽相同，主要可归纳为以下三种情况。

观点一：科技成果转化率是指已转化的科技成果占应用技术科技成果的比率。学者们认为"已转化的科技成果"并非指所有一切得到"转化"的科技成果。将应用技术成果用于生产并考察市场对该技术成果的可接受程度和直接利益或间接利益，若该应用技术成果可成功转化为商品并取得规模效益，则说明该项应用技术成果实现了转化。

观点二：科技成果转化率即已转化的科技成果占全部科技成果的比率。学者们认为，大多数的基础理论成果和部分软科学成果虽然无法直接应用于实际生产且成果转化的量化程度偏低，但其依然能够在一定程度上推动科技的进步与产业结构的调整和优化，因此建议将基础理论成果和软科学成果的转化情况纳入科技成果转化。

观点三：从管理角度来说，科技成果转化率应表示科技成果占全部研究课题的比率。

对于观点二来说，由于海洋领域的基础研究成果和软科学研究成果几乎都不能直接应用于生产实际，难以实现海洋科技成果的转化，因此不应采纳这一观点。对于观点三来说，定义中涉及的"科技成果"和"研究课题"来源于两套不同的海洋统计数据，其中"科技成果"来源于海洋科技统计数据，"研究课题"来源于海洋科技成果统计数据，因此这一观点不能正确地反映实际海洋科技成果转化情况。

因此，本报告建议采用观点一，对海洋科技成果转化率进行定义如下：海洋科技成果转化率是指一定时期内涉海单位进行自我转化或转化生产，处于投入应用或生产状态，并达到成熟应用的海洋科技成果占全部海洋科技应用技术成果的百分率。

根据海洋科技成果转化率的定义，可构建海洋科技成果转化率的公式为

海洋科技成果转化率=成熟应用的海洋科技成果/全部海洋科技应用技术成果×100%

由于海洋科技成果的转化是一个长期的过程，在测算海洋科技成果转化率时，覆盖周期越长，指标越符合实际。

需要注意的是，本报告所探讨的海洋科技成果转化率是狭义上的指标，公式中"成熟应用的海洋科技成果"和"全部海洋科技应用技术成果"均来自于海洋科技成果登记数据。从广义上来说，海洋科研课题、专利、论文、奖励、标准、软件著作权都属于海洋科技成果，难以统计且相互之间存在交叉重叠；从海洋科技成果形成，到初步应用，再到形成产品，直至达到规模化、产业化阶段，都可以算作海洋科技成果转化过程，难以辨别衡量。

基于海洋科技成果统计数据，运用海洋科技成果转化率标准公式进行计算，可得出2016年我国海洋科技成果转化率约为50.0%。

根据科技成果登记表，可将应用技术成果分为三个阶段：初期阶段、中期阶段和成熟应用阶段。初期阶段指实验室、小试等初期阶段的研究成果。中期阶段指新产品、新工艺、新生产过程直接用于生产前，为从技术上进一步改进产品、工艺或生产过程而进行的中间试验（中试）；为进行产品定型设计，获取生产所需技术参数而制备的样机、试样；为广泛推广而作的示范；为达到成熟应用阶段、广泛推广而进行的阶段性研究成果。成熟应用阶段指工业化生产、正式（或可正式）投入应用的成果，包括农业技术大面积推广，医疗卫生的临床应用，公安、军工的正样、定型等成果。

附录七 区域分类依据及相关概念界定

一、沿海省（自治区、直辖市）

拥有海岸线的11个省（自治区、直辖市），具体包括天津、河北、辽宁、上海、江苏、浙江、福建、山东、广东、广西和海南。

二、海洋经济区

我国有五大海洋经济区，分别为：环渤海经济区、长江三角洲经济区、海峡西岸经济区、珠江

三角洲经济区和环北部湾经济区。其中环渤海经济区中纳入评价的沿海省(直辖市)为辽宁、河北、山东、天津；长江三角洲经济区中纳入评价的沿海省(直辖市)为江苏、上海、浙江；海峡西岸经济区中纳入评价的沿海省为福建；珠江三角洲经济区中纳入评价的沿海省为广东；环北部湾经济区中纳入评价的沿海省(自治区)为广西和海南。

三、海洋经济圈

海洋经济圈分区依据《全国海洋经济发展"十二五"规划》，分别为北部海洋经济圈、东部海洋经济圈和南部海洋经济圈。北部海洋经济圈由辽东半岛、渤海湾和山东半岛沿岸及海域组成，本报告纳入评价的沿海省(直辖市)包括天津、河北、辽宁和山东；东部海洋经济圈由江苏、上海、浙江沿岸及海域组成，纳入评价的沿海省(直辖市)包括江苏、浙江和上海；南部海洋经济圈由福建、珠江口及其两翼、北部湾、海南岛沿岸及海域组成，即纳入评价的沿海省(自治区)包括福建、广东、广西和海南。

附录八 主要涉海高等学校清单(含涉海比例系数)

一、教育部直属高等学校

北京大学(0.0932，根据北京大学的涉海专业数占专业总数的比例确定涉海比例系数，下同)、清华大学(0.0256)、北京师范大学(0.1373)、中国地质大学(北京)(0.2381)、天津大学(0.0877)、大连理工大学(0.0886)、上海交通大学(0.0484)、南京大学(0.1163)、河海大学(0.9020)、浙江大学(0.1102)、厦门大学(0.0707)、中国海洋大学(0.8462)、武汉大学(0.0645)、中国地质大学(武汉)(0.2258)、中山大学(0.1280)、同济大学(0.0859)、华东师范大学(0.0789)、华中科技大学(0.0566)、华南理工大学(0.0490)。

二、工业和信息化部直属高等学校

哈尔滨工业大学(0.0462)。

三、交通运输部直属高等学校

大连海事大学(0.9348)。

四、地方高等学校

上海海洋大学(0.3191)、广东海洋大学(0.2200)、大连海洋大学(0.9545)、浙江海洋大学(0.8913)、宁波大学(0.1935)、集美大学(0.2388)、南京信息工程大学(0.2759)、海南热带海洋学院(0.1964)。

附录九 涉海学科清单(教育部学科分类)

附表 9-1 涉海学科清单(教育部学科分类)

代码	学科名称	说明
140	物理学	
14020	声学	
1402050	水声和海洋声学	原名为"水声学"

续表

代码	学科名称	说明
14030	光学	
1403064	海洋光学	
170	地球科学	
17050	地质学	
1705077	石油与天然气地质学	含天然气水合物地质学
17060	海洋科学	
1706010	海洋物理学	
1706015	海洋化学	
1706020	海洋地球物理学	
1706025	海洋气象学	
1706030	海洋地质学	
1706035	物理海洋学	
1706040	海洋生物学	
1706045	海洋地理学和河口海岸学	原名为"河口、海岸学"
1706050	海洋调查与监测	
	海洋工程	见 41630
	海洋测绘学	见 42050
1706061	遥感海洋学	亦名卫星海洋学
1706065	海洋生态学	
1706070	环境海洋学	
1706075	海洋资源学	
1706080	极地科学	
1706099	海洋科学其他学科	
240	水产学	
24010	水产学基础学科	
2401010	水产化学	
2401020	水产地理学	
2401030	水产生物学	
2401033	水产遗传育种学	
2401036	水产动物医学	
2401040	水域生态学	
2401099	水产学基础学科其他学科	
24015	水产增殖学	
24020	水产养殖学	
24025	水产饲料学	
24030	水产保护学	
24035	捕捞学	
24040	水产品贮藏与加工	
24045	水产工程学	
24050	水产资源学	
24055	水产经济学	

续表

代码	学科名称	说明
24099	水产学其他学科	
340	军事医学与特种医学	
34020	特种医学	
3402020	潜水医学	
3402030	航海医学	
413	信息与系统科学相关工程与技术	
41330	信息技术系统性应用	
4133030	海洋信息技术	
416	自然科学相关工程与技术	
41630	海洋工程与技术	代码原为57050，原名为"海洋工程"
4163010	海洋工程结构与施工	代码原为5705010
4163015	海底矿产开发	代码原为5705020
4163020	海水资源利用	代码原为5705030
4163025	海洋环境工程	代码原为5705040
4163030	海岸工程	
4163035	近海工程	
4163040	深海工程	
4163045	海洋资源开发利用技术	包括海洋矿产资源、海水资源、海洋生物、海洋能开发技术等
4163050	海洋观测预报技术	包括海洋水下技术、海洋观测技术、海洋遥感技术、海洋预报预测技术等
4163055	海洋环境保护技术	
4163099	海洋工程与技术其他学科	代码原为5705099
420	测绘科学技术	
42050	海洋测绘	
4205010	海洋大地测量	
4205015	海洋重力测量	
4205020	海洋磁力测量	
4205025	海洋跃层测量	
4205030	海洋声速测量	
4205035	海道测量	
4205040	海底地形测量	
4205045	海图制图	
4205050	海洋工程测量	
4205099	海洋测绘其他学科	
480	能源科学技术	
48060	一次能源	
4806020	石油、天然气能	
4806030	水能	包括海洋能等
4806040	风能	
4806085	天然气水合物能	
490	核科学技术	

续表

代码	学科名称	说明
49050	核动力工程技术	
4905010	舰船核动力	
570	水利工程	
57010	水利工程基础学科	
5701020	河流与海岸动力学	
580	交通运输工程	
58040	水路运输	
5804010	航海技术与装备工程	原名为"航海学"
5804020	船舶通信与导航工程	原名为"导航建筑物与航标工程"
5804030	航道工程	
5804040	港口工程	
5804080	海事技术与装备工程	
58050	船舶、舰船工程	
610	环境科学技术及资源科学技术	
61020	环境学	
6102020	水体环境学	包括海洋环境学
620	安全科学技术	
62010	安全科学技术基础学科	
6201030	灾害学	包括灾害物理、灾害化学、灾害毒理等
780	考古学	
78060	专门考古	
7806070	水下考古	
790	经济学	
79049	资源经济学	
7904910	海洋资源经济学	
830	军事学	
83030	战役学	
8303020	海军战役学	
83035	战术学	
8303530	海军战术学	

说明：根据二级学科所包含的涉海学科(三级学科)数占其所包含的三级学科总数的比例确定二级学科涉海比例系数如下：声学(0.06)，光学(0.06)，地质学(0.04)，海洋科学(1)，水产学基础学科(1)，水产增殖学(1)，水产养殖学(1)，水产饲料学(1)，水产保护学(1)，捕捞学(1)，水产品贮藏与加工(1)，水产工程学(1)，水产资源学(1)，水产经济学(1)，水产学其他学科(1)，特种医学(0.33)，信息技术系统性应用(0.25)，海洋工程与技术(1)，海洋测绘(1)，一次能源(0.36)，核动力工程技术(0.20)，水利工程基础学科(0.25)，水路运输(0.56)，船舶、舰船工程(1)，环境学(0.17)，安全科学技术基础学科(0.17)，专门考古(0.11)，资源经济学(0.17)，战役学(0.17)，战术学(0.17)。

编制说明

为响应国家海洋创新战略，服务国家创新体系建设，自然资源部第一海洋研究所自 2006 年起着手开展海洋创新指标的测算工作，并于 2013 年正式启动国家海洋创新指数的研究工作。《国家海洋创新指数报告 2017~2018》是相关系列报告的第五期，现将有关情况说明如下。

一、需求分析

创新驱动发展已经成为我国的国家发展战略，《中共中央关于全面深化改革若干重大问题的决定》明确提出要"建设国家创新体系"。海洋创新是建设创新型国家的关键领域，也是国家创新体系的重要组成部分。探索构建国家海洋创新指数，评价我国国家海洋创新能力，对海洋强国的建设意义重大。国家海洋创新指数系列报告编制的必要性主要表现在以下 4 个方面。

（一）全面摸清我国海洋创新家底的迫切需要

搜集海洋经济统计、科技统计和科技成果登记等海洋创新数据，全面摸清我国海洋创新家底，是客观分析我国国家海洋创新能力的基础。

（二）深入把握我国海洋创新发展趋势的客观需要

从海洋创新资源、海洋知识创造、海洋创新绩效和海洋创新环境 4 个方面，挖掘分析海洋创新数据，深入把握我国海洋创新发展趋势，以满足认清我国海洋创新路径与方式的客观需要。

（三）准确测算我国海洋创新重要指标的实际需要

对海洋科技进步贡献率、海洋科技成果转化率等海洋创新重要指标进行测算和预测，切实反映我国海洋创新的质量和效率，为我国海洋创新政策的制定提供系列重要指标支撑。

（四）全面了解国际海洋创新发展态势的现实需要

分析国际海洋创新发展态势，从海洋领域产出的论文与专利等方面分析国际海洋创新在基础研究和技术研发层面上的发展态势，全面了解国际海洋创新发展态势，为我国海洋创新发展提供参考。

二、编制依据

（一）十九大报告

党的十九大报告明确提出要"加快建设创新型国家"，并指出"创新是引领发展的第一动力，是建设现代化经济体系的战略支撑。要瞄准世界科技前沿，强化基础研究""加强国家创新体系建设，强化战略科技力量""坚持陆海统筹，加快建设海洋强国"。

（二）十八届五中全会报告

十八届五中全会报告指出，"必须把创新摆在国家发展全局的核心位置，不断推进理论创新、制度创新、科技创新、文化创新等各方面创新，让创新贯穿党和国家一切工作，让创新在全社会蔚然成风"。

(三)《国家创新驱动发展战略纲要》

中共中央、国务院 2016 年 5 月印发的《国家创新驱动发展战略纲要》指出,"党的十八大提出实施创新驱动发展战略,强调科技创新是提高社会生产力和综合国力的战略支撑,必须摆在国家发展全局的核心位置。这是中央在新的发展阶段确立的立足全局、面向全球、聚焦关键、带动整体的国家重大发展战略"。

(四)《中华人民共和国国民经济和社会发展第十三个五年规划纲要》

《中华人民共和国国民经济和社会发展第十三个五年规划纲要》提出创新驱动主要指标,强化科技创新引领作用,并指出"把发展基点放在创新上,以科技创新为核心,以人才发展为支撑,推动科技创新与大众创业万众创新有机结合,塑造更多依靠创新驱动、更多发挥先发优势的引领型发展"。

(五)《推动共建丝绸之路经济带和 21 世纪海上丝绸之路的愿景与行动》

《推动共建丝绸之路经济带和 21 世纪海上丝绸之路的愿景与行动》提出"创新开放型经济体制机制,加大科技创新力度,形成参与和引领国际合作竞争新优势,成为'一带一路'特别是 21 世纪海上丝绸之路建设的排头兵和主力军"的发展思路。

(六)《中共中央关于全面深化改革若干重大问题的决定》

《中共中央关于全面深化改革若干重大问题的决定》明确提出要"建设国家创新体系"。

(七)《"十三五"国家科技创新规划》

《"十三五"国家科技创新规划》提出"'十三五'时期是全面建成小康社会和进入创新型国家行列的决胜阶段,是深入实施创新驱动发展战略、全面深化科技体制改革的关键时期,必须认真贯彻落实党中央、国务院决策部署,面向全球、立足全局,深刻认识并准确把握经济发展新常态的新要求和国内外科技创新的新趋势,系统谋划创新发展新路径,以科技创新为引领开拓发展新境界,加速迈进创新型国家行列,加快建设世界科技强国"。

(八)《海洋科技创新总体规划》

《海洋科技创新总体规划》战略研究首次工作会上提出"要围绕'总体'和'创新'做好海洋战略研究""要认清创新路径和方式,评价好'家底'"。

(九)《"十三五"海洋领域科技创新专项规划》

《"十三五"海洋领域科技创新专项规划》明确提出"进一步建设完善国家海洋科技创新体系,提升我国海洋科技创新能力,显著增强科技创新对提高海洋产业发展的支撑作用"。

(十)《全国海洋经济发展规划纲要》

《全国海洋经济发展规划纲要》提出要"逐步把我国建设成为海洋强国"。

(十一)《全国科技兴海规划(2016—2020年)》

《全国科技兴海规划(2016—2020年)》提出，"到2020年，形成有利于创新驱动发展的科技兴海长效机制，构建起链式布局、优势互补、协同创新、集聚转化的海洋科技成果转移转化体系。海洋科技引领海洋生物医药与制品、海洋高端装备制造、海水淡化与综合利用等产业持续壮大的能力显著增强，培育海洋新材料、海洋环境保护、现代海洋服务等新兴产业的能力不断加强，支撑海洋综合管理和公益服务的能力明显提升。海洋科技成果转化率超过55%，海洋科技进步对海洋经济增长贡献率超过60%，发明专利拥有量年均增速达到20%，海洋高端装备自给率达到50%。基本形成海洋经济和海洋事业互动互进、融合发展的局面，为海洋强国建设和我国进入创新型国家行列奠定坚实基础"。

(十二)《国家中长期科学和技术发展规划纲要(2006—2020年)》

《国家中长期科学和技术发展规划纲要(2006—2020年)》提出，"要把提高自主创新能力摆在全部科技工作的突出位置""必须把提高自主创新能力作为国家战略，贯彻到现代化建设的各个方面，贯彻到各个产业、行业和地区，大幅度提高国家竞争力"，并指出科技工作的指导方针是"自主创新，重点跨越，支撑发展，引领未来"，强调要"全面推进中国特色国家创新体系建设"。

三、数据来源

《国家海洋创新指数报告2017～2018》所用数据来源如下。
(1)《中国统计年鉴》。
(2)《中国海洋统计年鉴》。
(3)科学技术部科技统计数据。
(4)教育部涉海高等学校和涉海学科科技统计数据。
(5)中国科学院兰州文献情报中心海洋科学论文、海洋专利等数据。
(6)中国科学引文数据库(Chinese Science Citation Database，CSCD)。
(7)科学引文索引扩展版(Science Citation Index Expanded，SCIE)数据库。
(8)德温特专利索引(Derwent Innovations Index，DII)数据库。
(9)工程索引(Engineering Index，EI)。
(10)海洋科技成果登记数据。
(11)《高等学校科技统计资料汇编》。
(12)其他公开出版物。

四、编制过程

《国家海洋创新指数报告2017～2018》受原国家海洋局科学技术司委托，由自然资源部第一海洋研究所海洋政策研究中心组织编写；中国科学院兰州文献情报中心参与编写了海洋论文、专利和国际海洋科技研究态势专题分析等部分；原国家海洋局科学技术司提供了我国海洋经济创新发展区域示范专题相关内容；科学技术部战略规划司、教育部科学技术司、国家海洋信息中心和华中科技大学管理学院等单位、部门提供了数据支持。编制过程分为前期准备阶段、数据测算与报告编制完善阶段、报告评审与修改完善阶段3个阶段，具体介绍如下。

(一)前期准备阶段

形成基本思路。2018年1~2月。国家海洋创新指数评价系列报告第一期(《国家海洋创新指数试评估报告 2013》)、第二期(《国家海洋创新指数试评估报告 2014》)、第三期(《国家海洋创新指数报告 2015》)、第四期(《国家海洋创新指数报告 2016》)分别于 2015 年 5 月、2015 年 12 月、2016 年 12 月和 2018 年 1 月出版。2018 年初,在《国家海洋创新指数报告 2017~2018》前期工作的基础上,经过多次研究讨论和交流沟通,总结归纳前四期的经验和不足之处,形成《国家海洋创新指数报告 2017~2018》的编制思路,编写《国家海洋创新指数报告 2017~2018》具体方案,汇报至原国家海洋局科学技术司。

收集数据。2018 年 1 月,顺利从华中科技大学科技统计信息中心和教育部科学技术司获取海洋科研机构科技创新数据、《高等学校科技统计资料汇编》相关数据和涉海高等学校按照涉海学科(一级)提取的涉海科技创新数据。同时,与中国科学院兰州文献情报中心合作,获取海洋领域 SCI 论文和海洋专利等数据。

组建报告编写组与指标测算组。2018 年 1 月,在《国家海洋创新指数报告 2016》原编写组的基础上,组建《国家海洋创新指数报告 2017~2018》编写组与指标测算组,具体由自然资源部第一海洋研究所海洋政策研究中心与中国科学院兰州文献情报中心等人员组成。

(二)数据测算与报告编制完善阶段

数据处理与分析。2018 年 1~2 月,对海洋科研机构科技创新数据及《中国统计年鉴》《中国海洋统计年鉴》《高等学校科技统计资料汇编》、涉海高等学校按照涉海学科(一级)提取的涉海科技创新数据等来源数据,进行数据处理与分析。

数据测算。2018 年 2 月 20 日至 3 月 20 日,测算海洋科技进步贡献率和海洋科技成果转化率,并根据相应的评价方法测算国家海洋创新指数和区域海洋创新指数。

报告文本初稿编写。2018 年 3 月 21 日至 4 月 20 日,根据数据分析结果和指标测算结果,完成报告第一稿的编写。

数据第一轮复核。2018 年 4 月 21 日至 5 月 7 日,组织测算组进行数据第一轮复核,重点检查数据来源、数据处理过程与图表。

报告文本第二稿修改。2018 年 5 月 8~22 日,根据数据复核结果和指标测算结果,修改报告初稿,形成征求意见文本第二稿。

数据第二轮复核。2018 年 5 月 23~31 日,组织测算组进行数据第二轮复核,流程按照逆向复核的方式,根据文本内容依次检查图表、数据处理过程、数据来源。

小范围征求意见。2018 年 6 月 1~8 日,进行小范围内部征求意见。

数据第三轮复核。2018 年 6 月 1~5 日,按照顺向与逆向结合复核的方式,核对数据来源、数据处理过程与文本图表对应。

报告文本第三稿完善。2018 年 6 月 9~14 日,根据数据第三轮复核结果和小范围征求意见情况,完善报告文本,形成征求意见第三稿。

(三)报告评审与修改完善阶段

根据专家咨询意见修改。2018 年 6 月 15 日,召开专家咨询会议,向专家汇报并根据专家意见修改文本。

内审及报告文本第四稿修改。2018 年 7 月 26 日至 8 月 2 日,中心组织进行内部审查,并根据意见修改文本。

管理部门审查。2018年7月26日至8月4日，报送自然资源部科技发展司[①]和科学技术部战略规划司审查，并根据意见修改文本。

计算过程复核。2018年7月30至8月4日，组织测算组进行计算过程的认真复核，重点检查计算过程的公式、参数和结果准确性，并根据复核结果进一步完善文本，结合各轮修改意见，形成征求意见第四稿。

顾问组审查。2018年8月，组织顾问组审查，并根据审查意见修改文本。

编写组文本校对。2018年9月1~25日，编写组成员按照章节对报告文本进行校对，根据各成员意见和建议修改完善文本。

出版社预审。2018年9月，向科学出版社提交文本电子版进行预审。

管理部门审核。2018年10月19日，报送自然资源部科技发展司审核。

五、意见与建议吸收情况

已征求意见30多人次。经汇总，收到意见和建议300多条。

根据反馈的意见和建议，共吸收意见和建议220多条。反馈意见和建议吸收率约为73.3%。

[①] 原国家海洋局科学技术司

更 新 说 明

一、增减了部分章节和内容

(1) 新增了第五章"我国海洋科研机构的空间分布特征与演化趋势"和第六章"全球海洋创新能力分析"。

(2) 删减了《国家海洋创新指数报告 2016》第六章"我国海洋经济创新发展区域示范专题分析"和第七章"我国海洋科技投入产出效率专题分析"的相应内容。

二、更新了国内和国际数据

(1) 更新了国际涉海创新论文数据。原始数据更新至 2016 年,用于海洋创新产出成果部分的分析,以及国内外海洋创新论文方面的比较分析。

(2) 更新了国际涉海专利数据。原始数据更新至 2016 年,用于海洋创新产出成果部分的分析,以及国内外海洋创新专利方面的比较分析。

(3) 更新了国内数据。国家海洋创新评价指标所用原始数据更新至 2016 年,区域海洋创新指数评价指标更新为 2016 年数据。

(4) 更新了数据来源。由于提供的企业数据的数量低于往年收集数量,相应数据的测算与比较受到较大影响,故剔除企业数据,形成新的原始数据,重新测算的各指数与以往报告中的数据会有相应差距。

三、调整了国家海洋创新指数指标体系

因 2016 年科学技术部科技统计数据中不再包含企业数据,对国家海洋创新指数指标体系中的分指数进行调整,删减海洋企业创新分指数。